# CHEMICAL WARFARE WITH SPECIAL REFERENCE TO NERVE GAS

## ARIN BHATTACHARYA

© Copyright is with the author

ISBN:- 978-1-300-88664-8

## DEDICATION

This dissertation is dedicated

to

My respected parents


I also thank god, my fellow mates

&

All the lovers of pharmacy world.

## Acknowledgement

In any of scientific literally work many people work towards achievement of the desired goal i.e. publication of the research work in form book. I would first like to thank Mr. Prashant Tiwari, M.Pharma (Pharmacology) for being the part of his ambitious project. Then I want to thanks my parents without their blessings, support and vision I could not have reached this stage.

Friendship is a treasured gift and fine friends are very few, My heartful thanks to my friend s who are always there for their support and positive views.

I would like to thanks the persons whom had criticize this project because it provided the positive impetus for completing the project in a more comprehensive and positive way.

Lastly it is only when someone writes a book that one realizes the true power of MSOffice software. It is simple- without this software this book would not be written.

Thank you Mr. Bill Gates and Microsoft Corp!

I strongly believe, "Success is blend of hard work and Destiny" and I think 'God' who have perpetually patronized me with consciousness and love to ladder the success. Thankful I ever remain....

*Arin Bhattacharya*

# Table of Contents

| CHAPTER NO. | NAME OF CHAPTER | PAGE NO |
|---|---|---|
| 1 | Introduction | 3 |
| 2 | History of chemical warfare | 6 |
| 3 | Introduction to nerve gas | 15 |
| 3.1 | Classification of nerve gas | 15 |
| 3.1.1 | G series | 15 |
| 3.1.2 | V series | 16 |
| 4 | Chemistry and property of nerve agents | 21 |
| 4.1 | Nerve agent VX | 21 |
| 4.2 | Nerve agent GA(Tabun) | 23 |
| 4.3 | Nerve agent GB(Sarin) | 25 |
| 5 | Mechanism of action of nerve gas | 30 |
| 6 | Clinical and Pharmacological effects of nerve gas | 33 |
| 6.1 | Signs and symptoms | 33 |
| 6.2 | Effects of nerve agent vapour | 33 |

| | | |
|---|---|---|
| 6.2.1 | Absorption | 33 |
| 6.2.2 | Local ocular effects | 33 |
| 6.2.3 | Local respiratory effects | 34 |
| 6.3 | Effects of liquid nerve agent | 34 |
| 6.3.1 | Local ocular effects | 34 |
| 6.3.2 | Local skin effects | 35 |
| 6.3.3 | Local gastrointestinal effects | 35 |
| 6.4 | Systemic effects of nerve agents | 35 |
| 6.4.1 | Poisoning | 35 |
| 6.4.1.1 | Muscarinic effects | 36 |
| 6.4.1.2 | Nicotinic effects | 37 |
| 6.4.1.3 | Central nervous system | 38 |
| 6.4.1.4 | Cumulative effects of repeated exposure | 39 |
| 6.5 | Potentially long term effects | 40 |
| 6.6 | Cause of death | 40 |
| 7 | Health effects of low exposure to nerve gases | 42 |
| 8 | Diagnosis and detection of nerve agents | 50 |
| 8.1 | Diagnosis of nerve agents | 50 |
| 8.2 | Detection of nerve agents | 50 |

| | | |
|---|---|---|
| 9.3.1.5 | Other anti cholinergics | 59 |
| 9.4 | Oximes | 59 |
| 9.4.1 | Toxicity | 59 |
| 9.5 | Enzyme reactivation | 60 |
| 9.6 | Anti convulsants | 61 |
| 9.7 | Respiratory supportive care | 61 |
| 9.7.1 | Air way management | 61 |
| 9.7.2 | Respiratory support | 61 |
| 9.7.3 | Positive pressure ventilation | 62 |
| 9.7.3.1 | Oxygen | 62 |
| 9.7.3.2 | General supportive care | 62 |
| 9.8 | Emergency field therapy | 62 |
| 9.8.1 | Self aid(or Buddy aid) | 63 |
| 9.8.2 | First aid by trained personnel | 63 |

| | | |
|---|---|---|
| 9.8.3 | Assisted ventilation | 64 |
| 10 | Prognosis of nerve gas exposure | 66 |
| 11 | Infamous sarin attack of the Tokyo subway | 67 |
| 11.1 | Background | 67 |
| 11.2 | Attack | 68 |
| 12 | Future prospects for improvement regarding nerve gas attack & prevention | 70 |
| 13 | Conclusion | 71 |
| 14. | References | 72 |

## Chapter 1 Introduction

We are surrounded by chemicals and have to use them in our day to day life. .But certain chemicals are there which could be used to create mass disaster and are major threat to human community. Aim of this literary highlights about the history of the chemical warfare, the current agents used in chemical warfare with special reference to nerve gas, chemistry of the nerve gases ,pharmacological, toxicological and theraputical parameters of these agents.

## CHAPTER 2 .HISTORY OF CHEMICAL WARFARE

Chemical warfare means using of the chemicals as a weapon against the enemy forces in the war. It is more lethal and devastating than the conventional method of warfare. Chemical agents were used as weapon from ancient time as old as the Stone Age. World war one and two, Gulf war had started the era of modern chemical warfare which is still continuing.

Poisons and incendiary weapons Greek historians first described in ancient myths included arrows dipped in serpent venom, water poisoned with drugs, plagues unleashed on armies, and secret formulas for combustible weapons. Exploiting lethal forces of nature was not just mythical fantasy, but was supported by numerous nonfiction authors in ancient times, to include Near Eastern records of 1770 BCE (before the Common Era), Greek myths recorded by Homer in about 750 BCE and from 500 BCE through the second century of the Common Era (CE). From 500 BCE on, weapons of poison and combustible chemicals in China and Japan were described in military and medical treatises. The development of Greek fire and other incendiaries was described in Byzantine and Islamic sources of late antiquity from the seventh century through the fourteenth century CE. Archers in antiquity created toxic projectiles with snake venoms, poisonous plants, and bacteriological substances.

Intoxicants were used to gain victories by ancient armies in Gaul, North Africa, Asia Minor, and Mesopotamia. The calmatives and intoxicants of antiquity included toxic honey, drugged sacrificial bulls, barrels of alcohol, and mandrake-laced wine. Malodors also had their origins in antiquity when, over two millennia ago, armies in Asia and Germany employed noxious smells to overwhelm their foes.

One of the earliest examples of chemical warfare was in the late Stone Age (10,000 BCE). Hunters known as the San, in southern Africa, used poison arrows. They dipped the wood, bone, and stone tips of their arrows in poisons obtained from scorpion and snake venoms, as well as poisonous plants (CBW Info, 2005; Tagate, 2006; Wikipedia, 2007a).[1,2]

In about 2000 BCE, soldiers in India used toxic fumes on the battlefield. Chinese writings from as far back as the seventh century BCE contain hundreds of recipes for the production of poisonous and irritating smokes for use in war, along with numerous accounts of their use. These accounts describe the arsenic containing ''soul-hunting fog'' and the use of finely divided lime dispersed into the air (Geiling, 2003; DeNoon, 2004;).[3,4]

The Assyrians in 600 BCE contaminated the water supply of their enemies by poisoning their wells with rye ergot (Mauroni, 2003)[5]. Solon of Athens used hellebore roots, a purgative that caused diarrhea, to poison the water in an aqueduct leading to the Pleistrus River, around 590 BCE, during the siege of Cirrha. The Cirrhaeans drank the water and developed violent and uncontrollable diarrhea, and were thus quickly defeated (Hemsley, 1987; United Kingdom Ministry of Defense, 1999; Noji, 2001; Robey, 2003; Tschanz, 2003;).[6-10]

In the Peloponnesian War between Athens and Sparta in the fifth century BCE, the Spartan forces used noxious smokes and flame unsuccessfully against the Athenian city of Plataea. Later the Boeotians successfully used noxious smokes and flame during the siege of Delium by placing a lighted mixture of coal, brimstone, and pitch at the end of a hollow wooden tube. Bellows pushed the resulting smoke through the tube and up to the walls of the besieged city, driving the defenders away (Thucydides, 1989).[11]

During the fourth century BCE, the Mohist sect in China used bellows to pump smoke from burning balls of mustard and other toxic vegetables into tunnels being dug by a besieging army (Tagate, 2006)[1]. In 200 BCE, the Carthaginians spiked wine with Mandrake root, a narcotic to sedate their enemies, feigned a retreat to allow the enemy to capture the wine, and then when the enemy was sleeping, returned to kill them. Around 50 CE, Nero eliminated his enemies with cherry laurel water that contained hydrocyanic acid (Hickman, 1999). [12]

Chinese soldiers during the period 960–1279 CE used arsenical smokes in battle (CNS, 2001)[13], and the Germans used noxious smokes in 1155 CE. In the fifteenth and sixteenth centuries, the Venetians used poison filled mortar shells and poison chests to taint wells, crops, and animals. During the Crimean War in 1854, British chemist Lyon Playfair proposed to fill a hollow shell with cacodyl cyanide [$(CH_3)_2AsCN$] for use against Russian ships (Miles, 1957b; Camerman and Trotter, 1963)[14-15]. John Doughty of New York City, a school teacher, sent a letter to the War Department suggesting that shells filled with chlorine be shot at the enemy during the American civil war.

World War I has been called the "Chemist's War" because it ushered in the beginning of the modern era of chemical warfare. Chlorine ($Cl_2$), designated Cl by the United States and Betholite by the French, is the only substance that has been used in its elementary state as a war gas. It is a greenish-yellow gas with an irritating and disagreeable odor. Chlorine causes spasm of the larynx muscles, burning of the eyes, nose, and throat, bronchitis, and asphyxiation. Its asphyxiating properties were first recognized by the Swedish chemist Karl Wilhelm Scheele in 1774. Chlorine was the first chemical agent used on a large scale by the Germans in April 1915 (Sartori, 1943; Field Manual 3–11.9, 2005).[16-17] Cyanogen chloride (CNCl) was discovered by Wurtz and first prepared by Berthollet in 1802.

It is a colorless gas with an irritating odor that immediately attacks the oral–nasal passages. Its symptoms are similar to hydrogen cyanide and in high concentrations, it eventually causes death.

The U.S. designation was CC which was later changed to CK. The French called it Mauguinite and Vitrite. It was first used by the French in October 1916. Phosgene ($COCl2$) or carbonyl chloride, designated CG by the United States, Collongite by the French, and D-Stoff by the Germans, was obtained in 1812 by Humphrey Davy when he exposed a mixture of chlorine and carbon monoxide to sunlight. Phosgene is a colorless gas with an odor like musty hay that attacks the lungs causing pulmonary edema and eventually death. It was first used by the Germans as a war gas in December 1915 (Sartori, 1943; Field Manual 3–11.9, 2005)[16-17].

Mustard agent or dichloroethyl sulfide ($S(CH2CH2)2Cl2$) was discovered by Despretz who obtained it by the reaction of ethylene on sulfur chloride in 1822. It is normally a pale yellow to dark brown oily liquid with odor like garlic (although the German mustard agent had an odor similar to mustard). The agent normally attacks the eyes and blisters the skin. The U.S. designated it HS, and then later HD after a purified version was developed in 1944. The French called it Yperite and the Germans Lost. The German name was derived by taking the first two letters of the two Germans Lommel and Steinkopf, who proposed and studied the use of this agent in warfare. The first use of mustard agent by the Germans near Ypres, Belgium, in July 1917, marked the beginning of a new phase of chemical warfare, and inflicted about 15,000 British casualties in three weeks.

Chloropicrin or trichloronitromethane ($CCl3NO2$) was prepared in 1848 by Stenhouse and was extensively used in World War I. It is a pungent, colorless, oily liquid that caused oral–nasal irritation, coughing, and vomiting. In high dosages, it causes lung damage and pulmonary edema. It was first employed dissolved in sulfuryl chloride by the Russians in 1916 in hand grenades. In Germany it was known as Klop, in France as Aquinite, and PS in the United States. It has also been used as an insecticide and fungicide, as well as for eradicating rats from ships (Sartori, 1943).[16]

At the beginning of World War I, both sides used munitions filled with irritants such as ethylbromoacetate ($CH_2BrCOOC_2H_5$), chloroacetone ($CH_3COCH_2Cl$, or French Tonite, German A-Stoff), o-dianisidine chlorosulfonate, xylyl bromide ($C_6H_4CH_3CH_2BR$, German T-Stoff), or benzyl bromide ($C_6H_5CH_2Br$) (Dogaroiu, 2003). Other irritants used in World War I included acrolein ($CH_2CHCHO$, French Papite), bromoacetone ($CH_3COCH_2Br$, U.S. BA, German B-Stoff, French Martonite), and bromobenzyl cyanide ($C_6H_5CHBrCN$, U.S. CA, British BBC, French Camite) (Salem et al., 2006). Thus, the first use of chemicals in World War I involved nonlethal tear gases, which were used by both the French and the Germans in late 1914 and early 1915.

Germany was the leader in first using chemical weapons on the battlefield and then introducing or developing new chemical agents to counter new developments in protective equipment. Fritz Haber was the designer behind many of Germany's chemical weapons. Although he was not a toxicologist, he profoundly influenced the science of chemical toxicology. Haber and colleagues conducted acute inhalation studies in animals with numerous chemical agents thought to be useful in chemical warfare. He also developed Haber's Law which is usually interpreted to mean that identical products of the concentration of an airborne agent and duration of exposure will yield similar biological responses (Witschi, 2000). Actually, the product of the concentration (C) of the gas in air in parts per million (ppm) and the duration of the exposure (t) in minutes was referred to as Haber's Constant (Haber, 1986). It was also referred to as the mortality-product, the Haber Product W ($C3t¼W$), or the lethal index, the lower the product or the index number, the greater the toxic power (Sartori, 1939). Although Haber's Law has been used by toxicologists to define acute inhalation toxicity for toxic chemicals, it can also be useful for quantitative risk or safety assessment (Rozman and Doull, 2001).[18]

Germany's use of chemical weapons on the battlefield began on October 27, 1914 when they fired shells loaded with dianisidine chlorosulfonate, a tear gas, at the British near Neuve Chapelle. This tear gas normally produces violent sneezing. In this case, however, the chemical dispersed so rapidly in the air that the British never knew they were attacked by gas (Charles, 2005).[19]

Cyanogen chloride (CNCl) was discovered by Wurtz and first prepared by Berthollet in 1802. It is a colorless gas with an irritating odor that immediately attacks the oral–nasal passages. Its symptoms are similar to hydrogen cyanide and in high concentrations, it eventually causes death. The U.S. designation was CC which was later changed to CK. The French called it Mauguinite and Vitrite. It was first used by the French in October 1916.

Phosgene ($COCl_2$) or carbonyl chloride, designated CG by the United States, Collongite by the French, and D-Stoff by the Germans, was obtained in 1812 by Humphrey Davy when he exposed a mixture of chlorine and carbon monoxide to sunlight. Phosgene is a colorless gas with an odor like musty hay that attacks the lungs causing pulmonary edema and eventually death. It was first used by the Germans as a war gas in December 1915.

Mustard agent or dichloroethyl sulfide ($S(CH_2CH_2)_2Cl_2$) was discovered by Despretz who obtained it by the reaction of ethylene on sulfur chloride in 1822. It is normally a pale yellow to dark brown oily liquid with odor like garlic (although the German mustard agent had an odor similar to mustard). The agent normally attacks the eyes and blisters the skin. The U.S. designated it HS, and then later HD after a purified version was developed in 1944. The French called it Yperite and the Germans Lost. The German name was derived by taking the first two letters of the two Germans Lommel and Steinkopf, who proposed and studied the use of this agent in warfare. The first use of mustard agent by the Germans near Ypres, Belgium, in July 1917, marked the beginning of a new phase of chemical warfare, and inflicted about 15,000 British casualties in three weeks.

Chloropicrin or trichloronitromethane (CCl3NO2) was prepared in 1848 by Stenhouse and was extensively used in World War I. It is a pungent, colorless, oily liquid that caused oral–nasal irritation, coughing, and vomiting. In high dosages, it causes lung damage and pulmonary edema. It was first employed dissolved in sulfuryl chloride by the Russians in 1916 in hand grenades. In Germany it was known as Klop, in France as Aquinite, and PS in the United States. It has also been used as an insecticide and fungicide, as well as for eradicating rats from ships (Sartori, 1943).

During the Rif War (1921–1926) in Spanish occupied Morocco, Spanish forces reportedly fired gas shells and dropped mustard agent bombs on the Riffians (SIPRI, 1971). [20]

In 1935, Italy used chemical weapons during their invasion of Ethiopia. The Italian military primarily dropped mustard agent in bombs, and experimentally sprayed it from airplanes and spread it in powdered form on the ground. In addition, there were reports that the Italians used and tear gas.

The history of nerve agents began on December 23, 1936 when Dr. Gerhard Schrader of I.G. Farben in Germany accidentally isolated ethyl N, N-dimethylphosphoramidocyanidate (C5H11N2O2P) while engaged in his program to develop new insecticides since 1934. It was a colorless to brown liquid with a faintly fruity odor.

Controlled animal laboratory studies revealed that death could occur within 20 min of exposure. In January 1937, Schrader and his assistant were the first to experience the effects on humans. A small drop spilled on a laboratory bench caused both of them to experience miosis and difficulty in breathing. Schrader reported the discovery to the Ministry of War which was required by the Nazi decree passed in 1935 that required all inventions of military significance be reported. The chemical was quickly recognized as a new, more deadly, chemical warfare agent. It was initially designated as Le-100, and later as Trilon-83. It would eventually become known as Tabun.

In 1938, Schrader discovered a second potent nerve agent, isopropyl methylphosphonofluoridate ($C_4H_{10}FO_2P$), whose name Sarin is an acronym for the names of the members of the development team: Schrader, Ambrose, Rudriger, and van der Linde. The Germans designated it T-144 or Trilon-46. The United States eventually designated it GB. It is a colorless liquid with no known odor. Animal tests indicated that it was 10 times more effective than Tabun

Soman (pinacolyl methyl phosphonofluoridate) ($C_7H_{16}FO_2P$), eventually designated GD by the United States, was developed by the Germans in 1944. Its name might have been either derived from the Greek verb ''to sleep'' or the Latin stem ''to bludgeon.'' It is a colorless liquid with a fruity or camphor odor. Soman was discovered by Dr. Richard Kuhn, a Nobel Laureate while he was working for the German Army on the pharmacology of Tabun and Sarin (SIPRI, 1971; MTS, 2005a; Field Manual 3–9.11, 2005). Soman combines features of both Sarin and Tabun (CBW). Initial tests showed that Soman was even more toxic than Tabun and Sarin (Harris and Paxman, 1982)[21]. The Germans also apparently researched two other nerve agents, later designated GE (ethyl Sarin, $C_5H_{12}FO_2P$) and GF (cycloSarin, $C_7H_{14}FO_2P$).

First identified in the 1930s by Kyle Ward Jr., eventually three nitrogen mustard agents were identified during the war. HN-1 ($C_6H_{13}Cl_2N$), HN-2 ($C_5H_{11}Cl_2N$), and HN-3 ($C_6H_{12}Cl_3N$) were all similar to mustard agent but had quicker reaction in the eyes.

In 1952.a new nerve agent was discovered by the British. The new agent, eventually designated VX (V for venomous), was developed at the ICI Protection Laboratory. VX ($C_{11}H_{26}NO_2PS$), both an organophosphate and an organosulfate compound which was immediately toxic to mammals as well as to insects was discovered by the British chemist Dr. Ranajit Ghosh. Its chemical name is O-ethyl S-[2 (diisopropylamino)ethyl] methylphosphonothiolate. VX is a colorless and odorless liquid.

The nerve agents are the most toxic of the known chemical warfare agents. They are hazards in their liquid and vapor states and can cause death within minutes after exposure. Nerve agents inhibit acetylcholinesterase in tissue, and their effects are caused by the resulting excess of acetylcholine.

Nerve agents are liquids under temperate condition. When dispersed, the more volatile ones constitute both a vapor and a liquid hazard. Others are less volatile and represent primarily a liquid hazard. The G-agents are more volatile than VX. GB (Sarin) is the most volatile, but evaporates less readily than water. GF is the least volatile of the G-agents (FAS). VX is persistent, that is, it does not degrade or wash away easily. The consistency of VX is similar to motor oil, so it is primarily a contact hazard.

The most significant use of chemical weapons occurred when Iraq used them against Iran during the Iran–Iraq War (1980–1988). The reports indicated extensive mustard agent and probable nerve agent usage.

In 1988, Iraq's military conducted a massive chemical agent attack by aircraft against their own people in Halabja, an unprotected city of 45,000 Iraqi Kurds, knowing they could not retaliate.

Libya reportedly used mustard agent against Chad in 1987 (Pringle, 1993). [22]

# CHAPTER 3. INTRODUCTION TO NERVE GAS

Nerve agents are a class of phosphorus containing organic chemicals (organophosphates) that disrupt the mechanism by which nerves transfer messages to organs. The disruption is caused by blocking acetylcholinesterase, an enzyme that normally relaxes the activity of acetylcholine, a neurotransmitter. Poisoning by a nerve agent leads to contraction of pupils, profuse salivation, convulsions, involuntary urination and defecation, and eventual death by asphyxiation as control is lost over respiratory muscles. Some nerve agents are readily vaporized or aerosolized and the primary portal of entry into the body is the respiratory system. Nerve agents can also be absorbed through the skin, requiring that those likely to be subjected to such agents wear a full body suit in addition to a respirator.

## 3.1 Classification of nerve gases

There are two main classes of nerve agents. The members of the two classes share similar properties and are given both a common name (such as *sarin*) and a two-character NATO identifier (such as GB).

### 3.1.1 G-Series

The *G-series* is thus named because German scientists first synthesized them. G series agents are known as Non-persistent, while the V series are persistent. All of the compounds in this class were discovered and synthesized during or soon after World War II, led by Dr. Gerhard Schrader (later under the employment of IG Farben).

This series is the first and oldest family of nerve agents. The first nerve agent ever synthesised was GA (tabun) in 1936. GB (sarin) was discovered next in 1939, followed by GD (soman) in 1944 and finally the more obscure GF (cyclosarin) in 1949.

**Figure 1 Structure of few G series agents**

**Tabun (GA, 1)**, **Sarin (GB, 2)**, **Soman (GD, 3)**, **Cyclosarin (GF, 4)**, **VX (5)**

## 3.1.2 V-Series

Dr. Ranajit Ghosh, a chemist at the Plant Protection Laboratories of Imperial Chemical Industries was investigating a class of organophosphate compounds (organophosphate esters of substituted aminoethanethiols). Like Dr. Schrader, an earlier investigator of organophosphates, Dr. Ghosh found that they were quite effective pesticides. In 1954, ICI put one of them on the market under the trade name Amiton. It was subsequently withdrawn, as it was too toxic for safe use. The toxicity did not go unnoticed and some of the more toxic materials had in fact been sent to the British Armed Forces research facility at Porton Down for evaluation. After the evaluation was complete, several members of this class of compounds would become a new group of nerve agents, the V agents (depending on the source, the V stands for Victory, Venomous, or Viscous). The best known of these is probably VX, with the Russian V-gas coming a close second (Amiton is largely forgotten as VG). This class of compounds is also sometimes known as Tammelin's esters, after Lars-Erik Tammelin of the Swedish Institute of Defense Research. Dr. Tammelin was also conducting research on this class of compounds in 1952, but for obvious reasons he did not publicize his work widely. The *V-series* is the second family of nerve agents and contains five well known members: VE, VG, VM, VR and VX, along with several more obscure analogues. The most studied agent in this family, VX, was invented in the 1950s at Porton Down in the United Kingdom. The other agents in this series have not been studied extensively and information about them is limited. It is known, however, that the V-series agents are about 10 times more toxic than the G-agent sarin (GB).

All of the V-agents are *persistent agents*, meaning that these agents do not degrade or wash away easily and can therefore remain on clothes and other surfaces for long periods. In use, this allows the V-agents to be used to blanket terrain to guide or curtail the movement of enemy ground forces. The consistency of these agents is similar to oil; as a result, the contact hazard for V-agents is primarily – but not exclusively – dermal. VX was the only V-series agent that was fielded by the US as a munition, consisting of rockets, artillery shells, airplane spray tanks and landmines.[9][10]

**Figure 2 Basic structure of V agents**

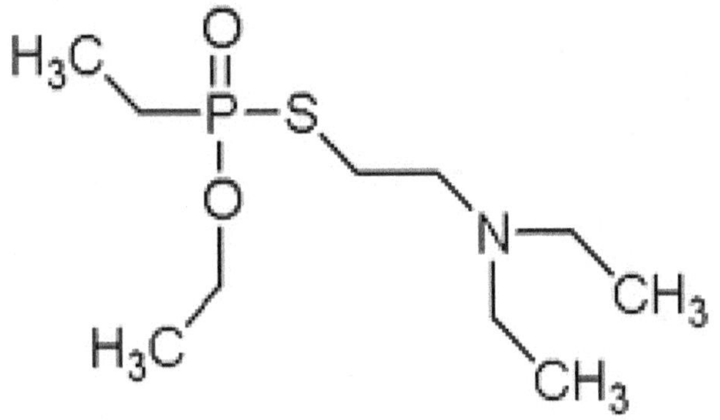

### 3.1.3 Novichok agents

The Novichok (Russian for "newcomer") agents are a series of organophosphate compounds that were developed in the Soviet Union from the mid 1960s to the 1990s. The goal of this program was to develop and manufacture highly deadly chemical weapons that were unknown to the West. These new agents were designed to be undetectable by standard NATO chemical detection equipment and to defeat chemical protective gear.

In addition to the newly developed "third generation" weapons, binary versions of several Soviet agents were developed and were designated as "Novichok" agents.

### Figure 3 Basic structure of Novichok" agents

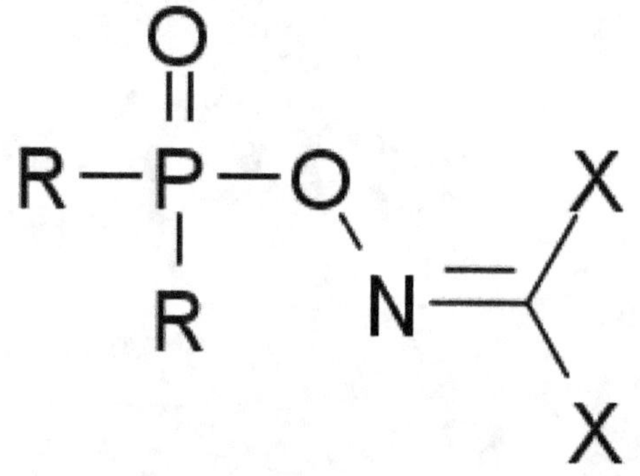

$$R-\underset{R}{\underset{|}{P}}(=O)-O-N=C(X)(X)$$

The first description of these agents was provided by Mirzayanov. Dispersed in an ultra-fine powder instead of a gas or a vapor, they have unique qualities. A binary agent was then created that would mimic the same properties but would either be manufactured using materials legal under the CWT or be undetectable by treaty regime inspections. The most potent compounds from this family, novichok-5 and novichok-7, are supposedly around 5-8x more potent than VX, however the exact structures of these compounds have not been reliably verified.

One of the key manufacturing sites was a chemical research institute in what is now Uzbekistan, and small, experimental batches of the weapons may have been tested on the nearby Ustyurt plateau.

Two broad families of organophosphorus agents have been claimed to be Novichok agents. First are a group of organophosphorus compounds with an attached dihaloformaldoxime group, with the general formula shown below, where R = alkyl, alkoxy, alkyl amino or fluorine and X = halogen (F, Cl, Br) or pseudohalogen such as C≡N. These compounds are extensively documented in Soviet literature of the time, but it is unclear whether they are in fact the potent "Novichok" compounds.

Some examples of the first group of compounds reported in the literature are shown below, but it is unknown whether any of these is novichok-5 or novichok-7.

Mirzayanov gives somewhat different structures for Novichok agents in his autobiography, as shown below. He makes clear that a large number of compounds were made, and many of the less potent derivatives reported in the open literature as new organophosphate insecticides, so that the secret chemical weapons program could be disguised as legitimate pesticide research.

# CHAPTER 4. CHEMISTRY AND PROPERTY OF SOME NERVE GASES

## 4.1 NERVE AGENTS—VX ($C_{11}H_{26}NO_2PS$)

**General Remarks:** Normally called VX, this nerve agent can be identified as (1) (O-ethyl S-[2-(diisopropylamino)ethyl)methylphosphonothioate; (2) O-ethyl S-[2-(diisopropylamino)ethyl) methylphosphonothiolate); or (3) methylphosphonothioic acid, S-[2-[bis(1-methylethyl)amino) ethyl O-ethyl ester. It bears CAS# 50782-69-9.

**Physical Properties**: With a molecular mass of 267.4, VX has mp _51$8$C, bp of 298$8$C, vapor pressure of 7310–4 mm Hg at 20$8$C, and a density similar to that of water (1.008 g=mL at 20$8$C).

It has a vapor density of 9.2 relative to air and is slightly soluble in water (approximately 30 g=L at room temperature). VX is both odorless and colorless.

## Synthesis

### Figure 4 Synthesis of VX nerve agents

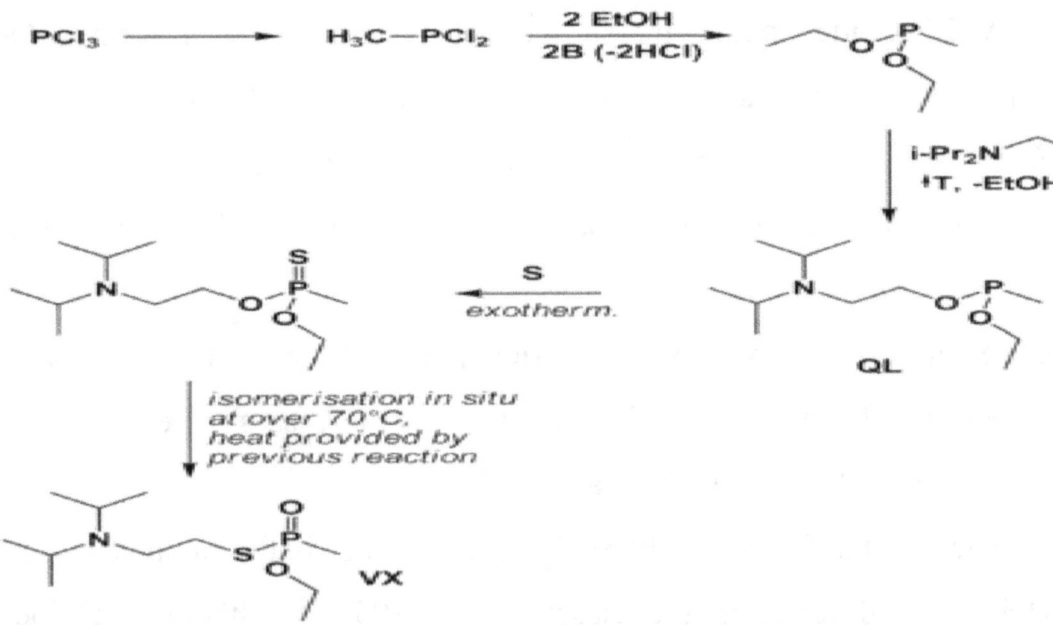

**Chemistry**: The most extensively studied of VX's reactions are related to its decomposition (oxidation and hydrolysis) and this, in turn, may be related to its persistence. Hydrolysis occurs rather slowly (at pH 7; 25°C; 1=2_30 days) and its products, of which there are several, are pH dependent. Thus P–S bond cleavage predominates at pH<6 or>10, but P–O cleavage is substantial in between. The former leads to 2-mercapto (N,N-diisopropyl)ethylamine, whereas the latter leads to S-(2-diisopropylaminoethyl) methyl phosphonothioate (Munro et al., 1999)[22]. Because nerve agents may hydrolyze over time (Yang et al., 1992, Wikipedia)[23-24] they have been stabilized for storage by moisture removal (Henderson, 2002)[25]

Oxidation is a potentially important route to destroying nerve agents. The oxidation reaction between VX and gaseous ozone has been shown to lead to a wide variety of products. In general, the most reasonable sites for initial oxidation of VX are sulfur or nitrogen, the former yielding a "sulfoxide," with the latter affording an amine oxide. Oxidation in a polar medium enhances S-oxide formation, perhaps (at least in part) by stabilizing the adjacent dipolar (S–O and P–O) bonds. It has been reported that VX was oxidized to "VX N-oxide" before the subsequent oxidation or hydrolysis to O-ethyl methylphosphonate using a variety of oxidants (Cassagne et al., 2001).[26]

The photochemical oxidation of both VX and sulfur mustard using $TiO_2$ or acetonitrile has been examined. Although effective, the reported quantum yields are somewhat low (<0.3%)

Highly reactive, nanosize magnesium oxide offers potential for the destruction of a VX, as well as GD and sulfur mustard. Half-lives for these processes were determined, as were decomposition products (Wagner et al., 1999).[27]

It has been suggested that Cu(II) catalyzes the hydrolysis of both thiophosphoric esters and methylphosphonofluoridates (Ketelaar et al., 1956)[28], although it does not significantly catalyze VX hydrolysis (Albizo and Ward, 1988)[29]. VX hydrolysis products may include a substance, EA2192, that is nearly as toxic as VX itself. The reactions of VX with a variety of reagents that should be capable of destroying it have been reviewed (Yang, 1999).[30] The fate of VX in the environment, including the influence of solids on hydrolysis, has been studied (Davisson et al., 2005)[31]. Supercritical water oxidation is of interest in the destruction of CW agents, including VX (Yesodharan, 2002).[32]

## 4.2 NERVE AGENT —GA (TABUN) ($C_5H_{11}N_2O_2P$)

**General Remarks**: Tabun and GA are the two most often used names for this nerve agent, although others include (1) O-ethyl N,N-dimethylphosphoramidocyanidate; (2) ethyl N,N-dimethylphosphoramidocyanidate; (3) Le-100; (4) N-Stoff; and (5) Trilon-83. It bears CAS# 77-81-6.

**Physical Properties**: Tabun is a colorless compound, which is said to have a somewhat fruity odor. The odor changes with decomposition so that with lesser amounts of decomposition the odor of cyanide (like bitter almonds) is apparent, whereas with greater amounts the odor of dimethylamine (like fish) is apparent. Tabun's vapor pressure is the lowest of all the G agents (approximately 0.04 mm Hg at 20$8$C). This suggests that tabun could be a relatively persistent threat at lower temperatures. Tabun has a vapor density of 5.6 relative to air and a liquid density of 1.08 g=mL (25$8$C). Its mp is _50$8$C, and its bp is approximately 240$8$C. It is three times as water soluble (approximately 10% at 20$8$C) as is VX, and also soluble in typical organic solvents (e.g., ethanol, diethyl ether, and chloroform). As with other nerve agents, dissolution in inert solvents (e.g., diethyl ether) enhances tabun's stability.

**Chemistry**: After its discovery in 1936, tabun was produced in Germany on a large scale in 1942 (Borkin, 1997).[33] Although others exist, one synthesis begins with phosphorus trichloride and employs the Arbuzov reaction (Kosolapoff, 1950).[34]

GA's acidic hydrolysis initially produces $C_2H_5OP(OH)(O)CN$ and dimethylamine (as salt) (Benes, 1963).[35] The CN group is lost next, followed by the loss of the EtO fragment. The ultimate product of hydrolysis is phosphoric acid.

Base-induced hydrolysis (and direct water hydrolysis) of tabun has been studied (Sanchez et al., 1993; McNaughton and Brewer, 1994)[36-37]. Analyzed by GC–MS, a number of previously unrecognized materials have been observed (D'Agostino and Provost, 1992).[38] Larsson (1952)[39] has reported a spectrophotometric study of tabun hydrolysis.

Ultimately, depending on conditions (pH, reaction times, and so forth), tabun hydrolysis products may include: cyanide ion; hydrocyanic acid, the monoethyl ester of dimethylphosphoramidic acid, ethanol, dimethylamine, and phosphoric acid. Under acidic conditions the P–N bond is cleaved early, whereas under alkaline conditions the P–C bond cleaves much more readily than does P–N. Tabun has a half-life of about 7 h at pH 4–5, but this increases to about 8.5 h at pH 7. Tabun hydrolysis is less than 5% complete after 20 h at pH 8.5. Tabun has a stereogenic (chiral) phosphorus atom and exists as a pair of enantiomers. A gas chromatograph study of the enantiomers of tabun has been reported (Degenhardt et al., 1986).[40]

Separation was achieved through the use of bis[(1R)-3-(heptafluorobutyryl-camphorate)nickel(II).

This approach also separated stereoisomers of both sarin and soman. These authors also reported the stereospecific hydrolysis of racemic tabun using phosphorylphosphatases. They noted the species (mouse, rat, horse) dependence of the hydrolysis. Dilute solutions of tabun in inert solvents (e.g., carbon tetrachloride) exhibit optical stability for months at -25 º C.

Shift reagents have aided in the study of pmr spectra of tabun and other nerve agents (Van Den Berg et al., 1984).[41] Two-dimensional NMR (1H-31P) has been applied to samples containing tabun (Albaret et al., 1997).[42]

## 4.3 NERVE AGENTS—GB (SARIN) ($C_4H_{10}FO_2P$)

**General Remarks**: GB carries several names including (1) isopropoxymethylphosphoryl fluoride, (2) isopropylmethylphosphonofluoridate, (3) methylphosphonofluoridic acid, 1-methylethyl ester, (4) Trilon-46, (5) N-stoff, and (6) T-144. It has CAS# 107-44-8.

**Physical Properties**: This colorless and odorless liquid has a molecular mass of 140.1 and a density of approximately 1.1 g=mL (20°C). Its mp is -57 °C and its bp is approximately 147 degree Centigrade.

Sarin is the most volatile of the G agents (vapor pressure 2.10 mm Hg at 20°C). It has been reported that sarin will evaporate from a sandy surface in about 2 h at 10 degree Centigrade (Sidell et al., 1998).[43]

**Chemistry**: Like tabun, sarin was first prepared (1938) by Schrader's group at I.G. Farben. An early synthesis involved the use of methylphosphonic dichloride ($CH_3P(O)Cl_2$) as a starting material. An interesting route involves the reaction of a tetraalkoxysilane, $(RO)_4Si$, as a source of the alkoxy fragment in GB (Black and Harrison, 1996a).[44]

The preparation of both nonradioactive (Bryant et al., 1960)[45] and radioactive (at P) sarin has been reported (Reesor et al., 1960).[46] Optically active (o.a.) sarin has been prepared using the enantiomeric sodium salts of O-isopropyl methylphosphonothioic acid. It has been found that o.a. sarin racemizes in less than 1 day when stored at 25°C in a polarimeter tube. On the other hand, dilute solutions in some solvents exhibit greater optical stability (Boter et al., 1966).[47]

GB is both miscible with water and hygroscopic, and like GA, is subject to both basic and acidic hydrolysis. The initial step in these hydrolyses involves loss of fluoride. Hydrolysis of the C–F bond in both GB and GD (soman) is accelerated by the presence of bleach (hypochlorite ion), although the process is complex and both pH and concentration dependent (Epstein et al., 1956).[48]

The effect of micelle formation on the hypochlorite-induced decomposition of toxic esters of phosphorus has been examined (Dubey et al., 2002).[49] The presence of the calcium and magnesium cations in sea water appears to accelerate the hydrolysis (Demek et al., 1970).[50]

The reaction of sarin with hydrogen chloride has been reported and kinetics determined by H1 NMR imaging (Bard et al., 1970).[51] With rate constants determined at 25°C, 81.5°C, and 100°C, Arrhenius analysis led to a calculated activation energy of 17.8 kcal/mole. The base-induced hydrolysis of sarin analogs and tabun was studied by Larsson (1958b)[52] and the half-life of GA has been estimated to be 1.5 min at pH=11 at 25°C. Ultimately, and depending on conditions (pH, reaction times, and so forth), hydrolysis products may include fluoride ion (or hydrogen fluoride), the 1-methylethyl ester of methylphosphonic acid, methylphosphonic acid, and 2-propanol. (þ)-Sarin is reported to be a weaker inhibitor of acetylcholinesterase (AChE) than is the enantiomer. Racemic sarin is reported to have an LD50 that is twice that of the (−) enantiomer. The kinetics of the fluoride-induced racemization of sarin is available (Christen and Van Den Muysenberg, 1965).[53]

Both sarin and soman react with aqueous $KHSO_5$ to produce the corresponding phosphonic acids (Yang et al., 1992).[54] Valuable studies of the use of hypervalent iodine derivatives (below) to hydrolyze phosphorus esters have been reported (Moss et al., 1983, 1984, 1986; Katritzky et al., 1988).[55-58]

## Figure 5  Physical, Chemical, Toxicological properties of the principal nerve agents

Physical, chemical and toxicological properties of the principal nerve agents

| Property | Tabun (GA) | Sarin (GB) | Soman (GD) | VX |
|---|---|---|---|---|
| Molecular weight / (Da) | 162.3 | 140.1 | 182.2 | 267.4 |
| Boiling point / (°C) | 230 | 158 | 198 | 298 |
| Melting point / (°C) | -49 | -56 | -80 | -20 |
| Vapor pressure / mm Hg | 0.037 (20 °C) | 2.1 (20 °C) | 0.40 (25 °C) | 0.0007 (20 °C) |
| Vapor Density (relative to air) | 5.6 | 4.9 | 6.3 | 9.2 |
| Liquid Density / (g per mL at 25 °C) | 1.08 | 1.10 | 1.02 | 1.01 |
| Volatility / (mg per m$^3$ a 25 °C) | 610 | 22000 | 3900 | 10.5 |
| Solubility in water | 9.8 g per 100g (25 °C) | miscible | 2.1g per 100g (20 °C) | Miscible |
| Persistency on soil | 1-1.5 day (half-life) | 2-24 h (5 to 20 °C) | relatively persistent | 2-6 days |
| LCt$_{50}$ in humans / (mg min per m$^3$) | 400 | 100 | 50 | 10 |
| LD$_{50}$ in humans / (mg per 70 kg in human) | 1000 (percutaneous) | 1700 (percutaneous) | 350 (percutaneous) | 6-10 (percutaneous) |

## Figure 6 IR SPECTRA OF NERVE GASES (G series)

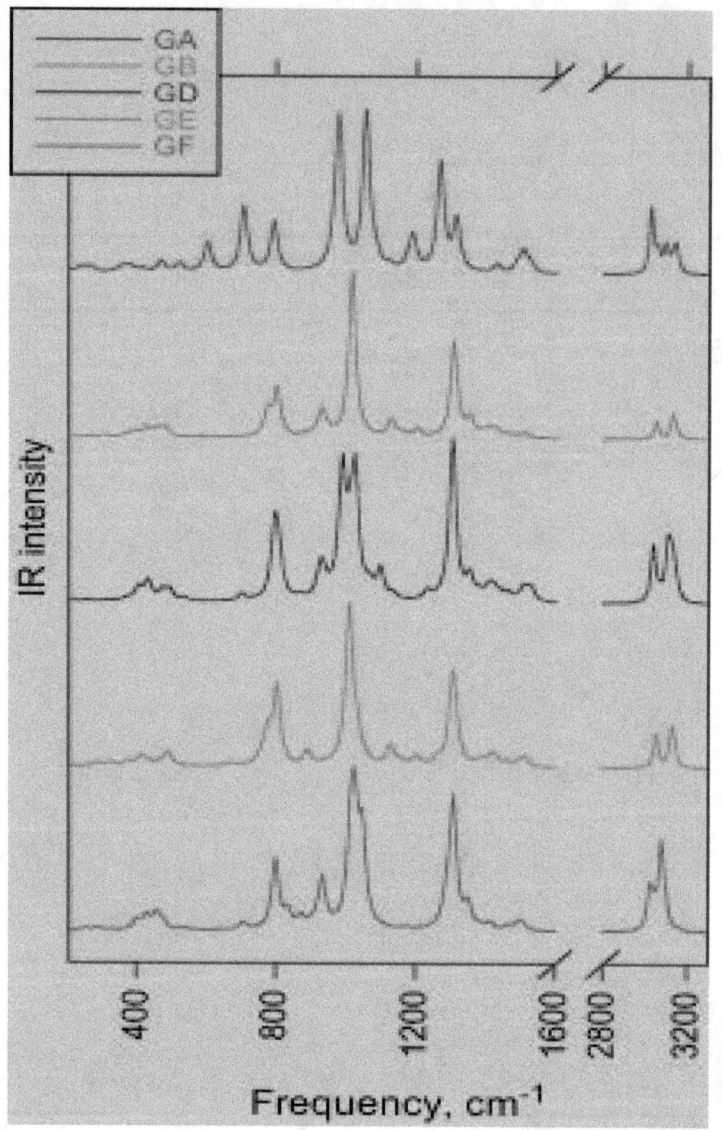

## Figure 7 MASS SPECTRA OF NERVE GASES

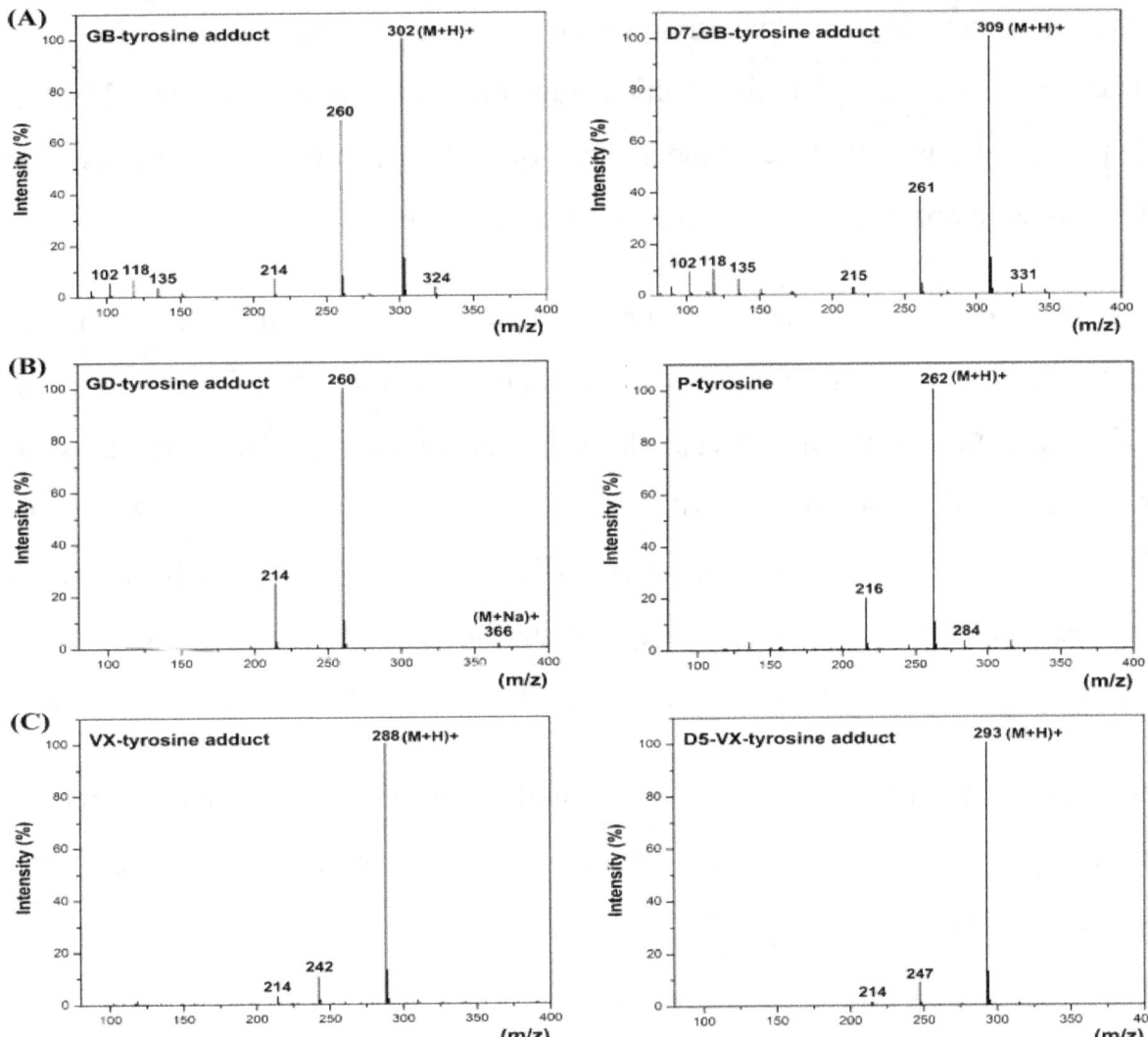

# CHAPTER 5   MECHANISM OF ACTION OF NERVE GAS

When a normally functioning motor nerve is stimulated it releases the neurotransmitter acetylcholine, which transmits the impulse to a muscle or organ. Once the impulse is sent, the enzyme acetylcholinesterase immediately breaks down the acetylcholine in order to allow the muscle or organ to relax.

Nerve agents disrupt the nervous system by inhibiting the function of acetylcholinesterase by forming a covalent bond with the site of the enzyme where acetylcholine normally undergoes hydrolysis (breaks down). The structures of the complexes of soman (one of the most toxic nerve agents) with acetylcholinesterase from *Torpedo californica* have been solved by X-ray crystallography. The result is that acetylcholine builds up and continues to act so that any nerve impulses are continually transmitted and muscle contractions do not stop.

This same action also occurs at the gland and organ levels, resulting in uncontrolled drooling, tearing of the eyes (lacrimation) and excess production of mucus from the nose (rhinorrhea).

# Figure 8 Flow chart of the mechanism of action of Nerve gas

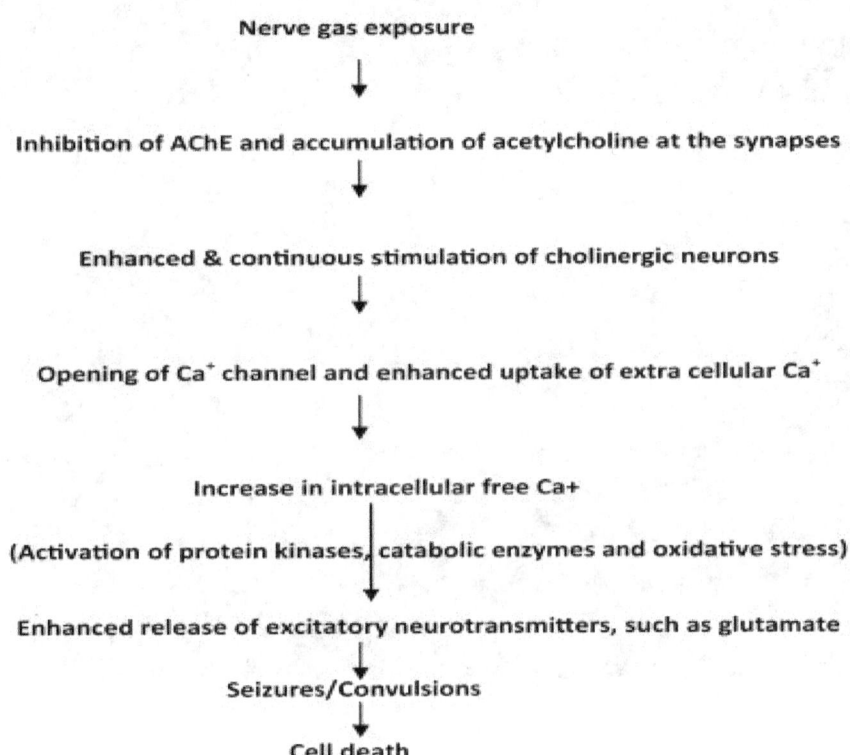

## Figure 9 Mechanism of action of Nerve gas

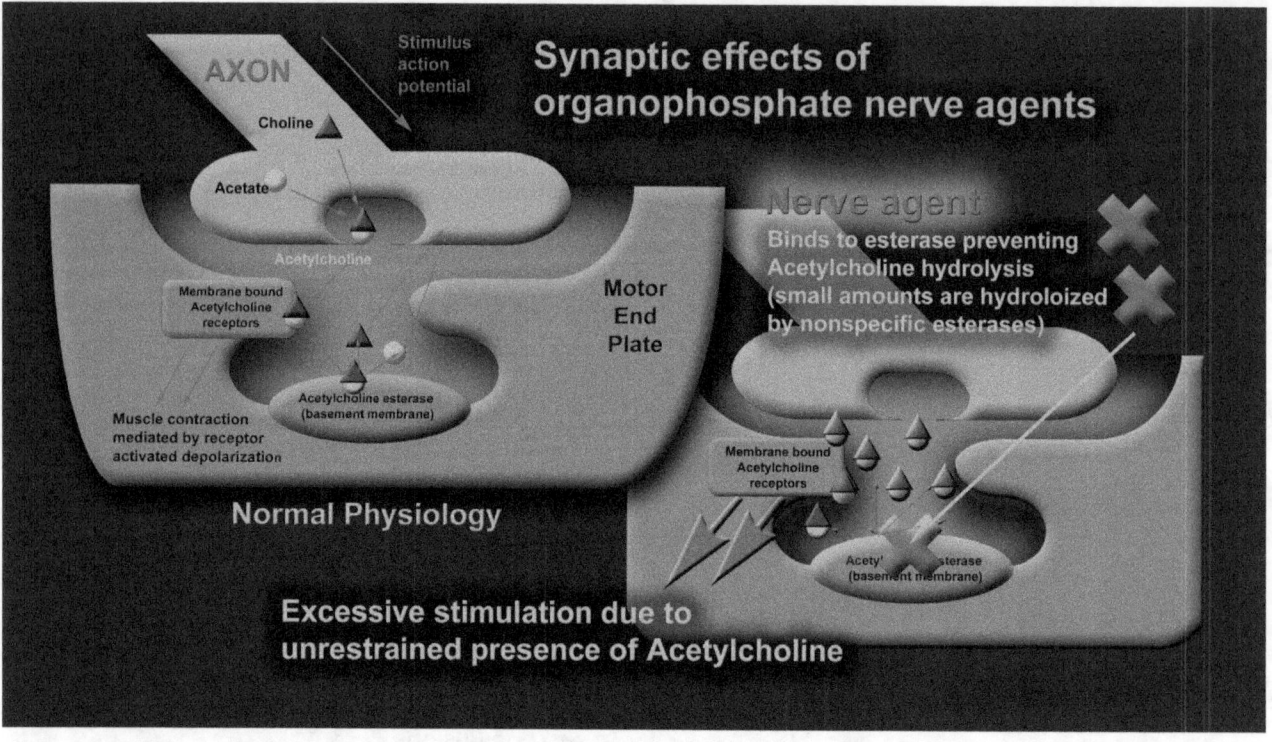

# CHAPTER 6 CLINICAL AND PHARMACOLOGICAL EFFECTS OF NERVE GASES

## 6.1 Signs and Symptoms

The order in which signs and symptoms appear and their relative severity depend on the route of exposure and whether the casualty has been exposed to liquid agent or vapour. The local effects of vapour and liquid exposure are described below, followed by a description of the systemic effects which occur after significant absorption of agent via any route.

## 6.2 Effects of Nerve Agent Vapour

### 6.2.1 Absorption

The lungs and the eyes absorb nerve agents rapidly. Changes occur in the smooth muscle of the eye (resulting in miosis) and in the smooth muscle and secretory glands of the bronchi (producing bronchial constriction and excessive secretions in the upper and lower airways). In high vapour concentrations, the nerve agent is carried from the lungs throughout the circulatory system; widespread systemic effects may appear in less than 1 minute.

### 6.2.2 Local Ocular Effects

These effects begin within seconds or minutes after exposure, before there is any evidence of systemic absorption. The earliest ocular effect which follows minimal symptomatic exposure to vapour is miosis. This is an invariable sign of ocular exposure to enough vapour to produce symptoms. It is also the last ocular manifestation to disappear. The papillary constriction may be different in each eye.

Within a few minutes after the onset of exposure, there also occurs redness of the eyes due to conjunctival hyperaemia, and a sensation of pressure with heaviness in and behind the eyes. Usually, vision is not grossly impaired, although there may be a slight dimness especially in the peripheral fields or when in dim or artificial light. Exposure to a level of a nerve agent vapour slightly above the minimal symptomatic dose results in miosis, pain in and behind the eyes attributable to ciliary spasm (especially on focusing), some difficulty of accommodation and frontal headache. The pain becomes worse when the casualty tries to focus the eyes or looks at a bright light. Some twitching of the eyelids may occur. Occasionally there is nausea and vomiting which, in the absence of systemic absorption, may be due to a reflex initiated by the ocular effects. These local effects may result in moderate discomfort and some loss of efficiency but may not necessarily produce casualties.

Following minimal symptomatic exposure, the miosis lasts from 24 to 72 hours. After exposure to at least the minimal symptomatic dose, miosis is well established within half an hour. Miosis remains marked during the first day after exposure and then diminishes gradually over 2 to 3 days after moderate exposure but may persist for as long as 14 days after severe exposure.

The conjunctival erythema, eye pain, and headache may last from 2 to 15 days depending on the dose.

### 6.2.3 Local Respiratory Effects

Following minimal exposure, the earliest effects on the respiratory tract are a watery nasal discharge, nasal hyperaemia, sensation of tightness in the chest and occasionally prolonged wheezing expiration suggestive of bronchoconstriction or increased bronchial secretion. The rhinorrhoea usually lasts for several hours after minimal exposure and for about 1 day after more severe exposure. The respiratory symptoms are usually intermittent for several hours duration after mild exposure and may last for 1 or 2 days after more severe exposure.

### 6.3 Effects of Liquid Nerve Agent

### 6.3.1 Local Ocular Effects

The local ocular effects are similar to the effects of nerve agent vapour. If the concentration of the liquid nerve agent contaminating the eye is high, the effects will be instantaneous and marked. If the exposure of the two eyes is unequal, the local manifestations may be unequal. Hyperaemia may occur but there is no immediate local inflammatory reaction such as may occur following ocular exposure to more irritating substances (for example, Lewisite).

### 6.3.2 Local Skin Effects

Following cutaneous exposure, there is localised sweating at and near the site of exposure and localised muscular twitching and fasciculation. However, these may not be noticed, causing the skin absorption to go undetected until systemic symptoms begin.

### 6.3.3 Local Gastrointestinal Effects

Following the ingestion of substances containing a nerve agent (which is essentially tasteless), the initial symptoms include abdominal cramps, nausea, vomiting and diarrhoea.

### 6.4 Systemic Effects of Nerve Agent

### 6.4.1 Poisoning

The sequence of symptoms varies with the route of exposure. Respiratory symptoms are generally the first to appear after inhalation of nerve agent vapour, although gastrointestinal symptoms are usually the first after ingestion. Following comparable degrees of exposure, respiratory manifestations are most severe after inhalation, and gastrointestinal symptoms may be most severe after ingestion. Otherwise, the systemic manifestations are, in general, similar after any exposure to nerve agent poisoning by any route. If local ocular exposure has not occurred, the ocular manifestations (including miosis) may initially be absent.

Following systemic absorption mydriasis may occur due to a nicotinic effect on the superior cervical ganglion.

The systemic effects may be considered to be nicotinic, muscarinic or central. The predominance of muscarinic, nicotinic or central nervous system effects will influence the amount of atropine, oxime or anticonvulsant which must be given as therapy.

Acute or delayed (up to 50 hours) pulmonary oedema has been described during the course of pesticide poisoning. Its exact mechanism is not known and may be due to a direct cardiac effect, haemodynamic modifications or both. Soman and other organophosphates have been reported to produce histopathological changes in the myocardium. Similar effects may thus occur with nerve agents.

### 6.4.1.1 Muscarinic Effects

Tightness in the chest is an early local symptom of respiratory exposure. This symptom progressively increases as the nerve agent is absorbed into the systemic circulation, whatever the route of exposure.

After moderate or severe exposure, excessive bronchial and upper airway secretions occur and may become very profuse, causing coughing, airway obstruction and respiratory distress. Audible wheezing may occur, with prolonged expiration and difficulty in moving air into and out of the lungs, due to the increased bronchial secretion or to bronchoconstriction, or both. Some pain may occur in the lower thorax and salivation increases.

Salivation may be so profuse that watery secretions run out of the sides of the mouth.

Bronchial secretion may be thick and tenacious. If postural drainage or suction is not employed, these secretions may add to the airway obstruction. Laryngeal spasm followed by collapse of the hypopharyngeal musculature may also obstruct the airway.

The casualty may gasp for breath, froth at the mouth, and become cyanotic. If the upper airway becomes obstructed by secretions, laryngeal spasm or hypopharyngeal musculature collapse, or if the bronchial tree becomes obstructed by secretions or bronchoconstriction, little ventilation may occur despite respiratory movements. As hypoxaemia and cyanosis increase, the casualty will fall exhausted and become unconscious.

Following inhalation of nerve agent vapour, the respiratory manifestations predominate over the other muscarinic effects; they are likely to be most severe in older casualties and in those with a history of respiratory disease, particularly bronchial asthma. However, if the exposure is not so overwhelming as to cause death within a few minutes, other muscarinic effects appear.

These include sweating, anorexia, nausea and epigastric and sub-sternal tightness with heartburn and eructation. If absorption of nerve agent has been great enough (whether due to a single large exposure or to repeated smaller exposures), there may follow abdominal cramps, increased peristalsis, vomiting, diarrhoea, tenesmus, increased lachrymation and urinary frequency.

Cardiovascular effects can include bradycardia, hypotension and cardiac arrhythmias. The casualty perspires profusely and may have involuntary defecation and urination; this may lead to circulatory collapse and cardiorespiratory arrest followed by death.

### 6,4.1.2 Nicotinic Effects

With the appearance of moderate muscarinic systemic effects, the casualty begins to have increased fatigue and mild generalized weakness which is increased by exertion. This is followed by involuntary muscular twitching, scattered muscular fasciculations and occasional muscle cramps. The skin may be pale due to vasoconstriction and blood pressure moderately elevated (transitory) together with a tachycardia, resulting from cholinergic stimulation of sympathetic ganglia and possibly from the release of epinephrine. If the exposure has been severe, the muscarinic cardiovascular symptoms will dominate and the fascicular twitching (which usually appear first in the eyelids and in the facial and calf muscles) becomes generalized.

Many rippling movements are seen under the skin and twitching movements appear in all parts of the body. This is followed by severe generalised muscular weakness, including the muscles of respiration. The respiratory movements become more laboured, shallow and rapid; then they become slow and finally intermittent. Later, respiratory muscle weakness may become profound and contribute to the respiratory depression.

### 6.4.1.3 Central Nervous System Effects

In mild exposures, the systemic manifestations of nerve agent poisoning usually include tension, anxiety, jitteriness, restlessness, emotional lability, and giddiness. There may be insomnia or excessive dreaming, occasionally with nightmares.

If the exposure is more marked, the following symptoms may be evident:

- Headache.

- Tremor.

- Drowsiness.

- Difficulty in concentration.

- Impairment of memory with slow recall of recent events.

- Slowing of reactions.

In some casualties there is apathy, withdrawal and depression. With the appearance of moderate symptoms, abnormalities of the EEG occur, characterised by irregularities in rhythm, variations in potential, and intermittent bursts of abnormally slow waves of elevated voltage similar to those seen in patients with epilepsy. These abnormal waves become more marked after one or more minutes of hyperventilation which, if prolonged, may occasionally precipitate a generalised convulsion.

If absorption of nerve agent has been great enough, the casualty becomes confused and ataxic. The casualty may have changes in speech, consisting of slurring, difficulty in forming words, and multiple repetition of the last syllable. The casualty may then become comatose, reflexes may disappear and respiration may become Cheyne-Stokes in character. Finally, generalised seizures may ensue, but not all of these may be associated with convulsive activity.

With the appearance of severe central nervous system symptoms, central respiratory depression will occur (adding to the respiratory embarrassment that may already be present) and may progress to respiratory arrest.

However, after severe exposure the casualty may lose consciousness and convulse within a minute without other obvious symptoms. Death is usually due to respiratory arrest and anoxia, and requires prompt initiation of assisted ventilation to prevent death. Depression of the circulatory centres may also occur, resulting in a marked reduction in heart rate with a fall of blood pressure some time before death.

### 6.4.1.4 Cumulative Effects of Repeated Exposure

Repeated exposure to concentrations of a nerve agent insufficient to produce symptoms following a single exposure, may result in the onset of symptoms. Continued exposure, possibly over several days, may be followed by increasingly severe effects.

After symptomatic exposure, increased susceptibility may persist for up to 3 months (the body synthesizes acetylcholinesterase at a rate of approximately 1% new enzyme activity per day). A period of increased susceptibility occurs during the enzyme regeneration phase which could last from weeks to months, depending on the severity of the initial exposure. During this period the effects of repeated exposures are cumulative.

Increased susceptibility is not limited to the particular nerve agent or other cholinesterase inhibitors initially absorbed.

## 6.5 Potential Long Term Effects

Minor EEG changes have been noted more than a year after a clinically relevant, symptomatic nerve agent exposure; averaged EEGs in a group of people who had been exposed to a nerve agent were compared to a control group. Since changes could not be identified in individual EEGs, the clinical relevance is unknown. Neuropsychiatric changes have been noted in individuals for weeks to months after insecticide poisoning.

Polyneuropathy, reported after organophosphate insecticide poisoning, has not been reported in humans exposed to nerve agents. It has been produced in animals only at doses of nerve agents so high that survival would be unlikely. The Intermediate Syndrome (polyneuropathy in the medium term – weeks) has not been reported in humans after nerve agent exposure, nor has it been produced in animals by nerve agent administration. Muscular necrosis has been produced in animals after high dose nerve agent exposure but resolves within weeks; it has not been reported in humans.

## 6.6 Cause of Death

In the absence of treatment, death is caused by asphyxia resulting from airway obstruction, paralysis of the muscles of respiration and central depression of respiration. Airway obstruction is due to pharyngeal muscular collapse, upper airway and bronchial secretions, bronchial constriction and occasionally laryngospasm and paralysis of the respiratory muscles.

Respiration is shallow, laboured, and rapid and the casualty may gasp and struggle for air. Cyanosis increases. Finally, respiration becomes slow and then ceases. Unconsciousness ensues. The blood pressure (which may have been transitorily elevated) falls.

Cardiac rhythm may become irregular and death may ensue. If assisted ventilation is initiated via cricothyroidotomy or endotracheal tube and airway secretions are removed (by postural drainage and suction, and diminished by the administration of atropine), the individual may survive several lethal doses of a nerve agent, albeit with a potential for severe irreversible brain damage. However, if the exposure has been overwhelming, amounting to many times the lethal dose, death may occur despite treatment as a result of respiratory arrest and cardiac arrhythmia.

When overwhelming doses of the agent are absorbed quickly, death occurs rapidly without orderly progression of symptoms equally timely administration of necessary therapeutic measures. Urgent skin decontamination and continued respiratory protection must be applied alongside the administration of autoinjectors for self and buddy aid, all of which must be maintained during rapid evacuation to medical care.

Triage priorities must be reviewed frequently, with subsequent delivery of appropriate drug therapy and continued respiratory support, where necessary. This must be administered in a timely and confident manner if emergency medical management is to be successful and the ultimate prognosis favourable. Field medical management may have to be conducted initially in a contaminated environment, requiring practised procedures to avoid cross-contamination and unnecessary exposure of the already compromised casualty.

# CHAPRTER 7 HEALTH EFFECTS OF LOW-LEVEL EXPOSURE TO NERVE AGENTS

The nerve agents are highly toxic organophosphorous (OP) compounds that are chemically related to some insecticides (parathion, malathion). The five most common nerve agents are tabun (o-ethyl N,N-dimethyl phosphoramidocyanidate; military designation GA), sarin (isopropyl methyl phosphonofluoridate; military designation = GB), soman (pinacolyl methyl phosphonofluoridate; military designation= GD), cyclosarin (cyclohexyl methylphosphonofluoridate; military designation = GF), and VX (o-ethyl S-2-N,N-diisopropylaminoethyl methyl phosphonofluoridate).

They exert their toxic effects by inhibiting the cholinesterase (ChE) family of enzymes to include acetylcholinesterase (AChE; E.C.3.1.1.7), a critically important central nervous system (CNS) and peripheral nervous system (PNS) enzyme that hydrolyzes the neurotransmitter acetylcholine (ACh).

Although the nerve agents can inhibit other esterases, their potency and specificity for inhibiting AChE account for their exceptionally high toxicity. For example, the rate constants for inhibition of AChE by soman, sarin, tabun, or VX are 2–3 orders of magnitude greater than for the more commonly known OP compounds such as DFP, paraoxon, or methylparaoxon (Gray and Dawson, 1987).[59] Likewise, the rate constants for inhibition of AChE by the nerve agents are also 2–5 times greater than for trypsin (E.C.3.4.21.4), chymotrypsin (E.C.3.4.21.1), or carboxylesterase (E.C.3.1.1.1) (Maxwell and Doctor, 1992)[60] indicative of selective inhibition of this enzyme.

Nerve agents bind to the active site of the AChE enzyme, thus preventing it from hydrolyzing ACh. The enzyme is inhibited irreversibly and the return of esterase activity depends on the synthesis of new enzyme (approximately 1%–3% per day in humans). All agents are highly lipophylic and readily penetrate the CNS. Acetylcholine is the neurotransmitter at the neuromuscular junction of skeletal muscle, the preganglionic nerves of the autonomic nervous system, the postganglionic parasympathetic nerves, as well as muscarinic and nicotinic cholinergic synapses within the CNS. Following nerve agent exposure and the inhibition of approximately >40% of the AChE enzyme pool, levels of ACh rapidly increase at the various effector sites resulting in continuous overstimulation. It is this hyperstimulation of the cholinergic system at central and peripheral sites that leads to the toxic signs of poisoning with these compounds. The signs of poisoning include miosis (constriction of the pupils), increased tracheobronchial secretions, bronchial constriction, laryngospasm, increased sweating, urinary and fecal incontinence, muscle fasciculations, tremor, convulsions=seizures of CNS origin, and loss of respiratory drive from the CNS.

The relative prominence and severity of a given sign are highly dependent on the route and degree of exposure. Ocular and respiratory effects occur rapidly and are most prominent following vapor exposure, whereas localized sweating, muscle fasciculations, and gastrointestinal disturbances are the predominant signs following percutaneous exposures and usually develop in a more protracted fashion. The acute lethal effects of the nerve agents are generally attributed to respiratory failure caused by a combination of effects at both central and peripheral levels and are further complicated by copious secretions, muscle fasciculations, and convulsions. There are several excellent reference sources that provide more detailed discussions of the history, chemistry, physiochemical properties, pharmacology, and toxicology of the nerve agents (Koelle, 1963; Sidell, 1992; Somani et al., 1992; Marrs et al., 1996; Taylor, 2001).[61,62,63,54,65]

Farahat et al. (2003)[66] evaluated the neurobehavioral performance on a battery of tests given to unexposed controls and Egyptian workers that applied pesticides (OPs, carbamates, insect growth regulators, pyrethroids) to cotton crops during the height of the application season. The pesticide applicators had slightly (20%) lowered serum ChE levels compared with the controls, but these levels were still within the normal range. The exposed group showed significantly poorer performance than the controls on six of the neurobehavioral battery tests (similarities, digit symbol substitution, trailmaking parts A and B, letter cancellation, digit span, and Benton visual retention). In addition, exposed subjects reported higher instances of the neurological symptoms of dizziness and numbness and significantly higher neuroticism (nervousness, anxiety) scores than controls. Serum ChE was not significantly correlated with task performance, but duration of pesticide exposure was. The authors suggested that the deficits in a wider array of neurobehavioral tests observed in this than in other studies, coupled with the neurological signs, indicated a higher level of exposure in this study population. Roldan-Tapia et al. (2005)[67] studied neurobehavioral performance of greenhouse workers exposed occupationally to both OPs and carbamate pesticides. Individuals with a high cumulative exposure risk ("years working with pesticides") displayed worsened perceptive function performance, visuomotor praxis, and integrative task performance times. These findings were taken as evidence that long-term use of these pesticides has adverse effects on neurobehavioral functioning. Rothlein et al. (2006)[68] reported that neurobehavioral performance of Hispanic farm workers with high risk for exposure to pesticides was lower than similar Hispanic immigrants in nonagricultural jobs. Within the exposed farm worker population there was a positive correlation between severity of exposure, as measured by urinary OP metabolite levels, and poorer performance on some of the neurobehavioral tests. This same group has also used a computerized test battery to examine whether adolescent farm workers exposed to pesticides are any more susceptible to these neurobehavioral effects than adults (Rohlman et al., 2006).[69] Although there was no evidence of selective sensitivity between adolescents and adults in this study, the results showed that cumulative exposure to low levels of pesticides is associated with neurological impairment as measured by tests of selective attention, digit symbol substitution, and reaction time.

In contrast, Eckerman et al. (2007)[70] have assessed neurobehavioral function in rural adolescents involve with agriculture and urban controls. When level of pesticide exposure was factored into a multiple linear regression analysis of performance, it was found that poorer performance on finger tapping, digit span, and selective attention was associated with high levels of exposure and that this relationship was especially strong for the youngest (10–11 years) age group.

Bazylewicz-Walczak et al. (1999)[71] published their study of greenhouse workers occupationally exposed to pesticides. They gave greenhouse workers and a matched control group neuropsychological tests (simple reaction time, digit symbol, digit span, Benton visual retention test, Santa Ana test, aiming test, POMS) and a subjective symptom questionnaire before and then 4 month s later after the heaviest period of OP pestici de application. Overall, when compared with the controls, the exposed subjects showed slower simple reaction times, lower hand movement efficiency on the aiming test, and reported a higher degree of anxiety, anger, depression, and fatigueinertia. In addition they also reported more complaints relating to absent-mindedness and neurological symptoms. There were no differences in the exposed group over the one season. There was no change or an improvement in scores on the neuropsychological tests across the season while there was an improvement in mood and general feeling scores between the pre- and the postseason tests. The authors concluded that even low, long-term OP pesticide exposure may be associated with subtle adverse behavioral effects, and that they are characterized by increased tension and anxiety states, depression, fatigue, and a slow-down of perceptual-motor functions atropinized guinea pigs. Animals were exposed to 200 ppb for 5 h of a toxic steroisomer of soman, which resulted in a gradual inhibition of RBC–AChE to approximately 10% of baseline. This level of exposure resulted in an insignificant reduction of AChE activity in brain and diaphragm although it was equivalent to Ct values of 0–48 mg min=m3, a dose well above one sufficient to cause an incapacitating miosis. The observed lack of inhibition of AChE in brain and diaphragm at the end of the long-term, low-level exposure was interpreted to mean that systemic intoxication is unlikely despite extensive inhibition of blood AChE. Furthermore, Benschop et al.[72] argued that the development of persistent neuropsychological disorders under these conditions would be unlikely. The authors cautioned that studies in animals without the benefit of carboxylesterase binding sites, such as primates, would most probably reflect a different outcome. This last study points out the influence of dose rate in determining whether a given exposure would be ''nonlethal,'' ''subtoxic,'' or ''subclinical,'' a point made as long ago as 1975 by Sim. The latter wrote that a patient appearing in a clinic without measurable ChE, yet not appearing to be intoxicated, ''emphasizes that the poison is cumulative and if taken into the body slowly, can be accommodated without the appearance of critical illness.''

In his brief review of chronic effects of low-level exposure to anti-ChEs. It was concluded that ''concerns about major adverse health effects of low-level exposure to anti-ChEs in general seem entirely unwarranted on the basis of currently available literature, but the data are at present insufficient to reflect the possibilities of subtle, agent-specific effects.'' Riddle et al. (2003)[73] in their review of potential Gulf War nerve agent exposures found (1) difficulties in documenting symptomatic exposures to nerve agents, (2) corresponding lack of specific exposure-related health effects, yet (3) commented that feelings of helplessness in face of an uncertain exposure could be overwhelming. A study by Atchison et al. (2004)[74] provided information on the initial validation of a guinea pig model of low-dose exposure to the nerve agents sarin, soman, and VX. This model was then used in a number of subsequent studies that evaluated physiological, electrophysiological, biochemical, and behavioral effects of this exposure paradigm.

In the Atchison et al. (2004)[74] study, male guinea pigs were injected (sc) once a day, 5 days a week (M–F), for 2, 4, or 13 weeks with 0.2, 0.4, 0.6, and 0.8 3acute LD50s previously established of sarin, soman, or VX. These exposure times (2, 4, and 13 weeks) were designed to cover the subacute to chronic durations of exposure as defined by Klaassen and Eaton (1991).[75] Signs of cholinergic toxicity (for 2 h after exposure) and body weights were monitored throughout the exposure period. After the initial study, a second study was performed using asymptomatic doses (vehicle, 0.2 and 0.43LD50) administered for 2 or 4 weeks according to the same protocol. In these animals, red-blood cell AChE was measured 1 h after the last injection, then 3 days later, the animals were sacrificed and diaphragm tissue was collected for measurement of AChE activity, blood was sampled for standard chemistries and hematology, and tissues (brain, heart, lung, liver, gallbladder, kidney, adrenal gland, skeletal muscle, sciatic nerve,

Health Effects of Low-Level Exposure to Nerve Agents 83 and testes) were collected for histopathology (hematoxylin and eosin stain) examination by a board certified veterinary pathologist. Doses of 0.2 or 0.43LD50 of any of the three agents were tolerated with no lethality or toxic signs throughout the 13 week exposure. In contrast, 0.83LD50 of the three agents were lethal to all animals exposed to this dose, and 0.63LD50 of sarin was also lethal while these same dose fractions of soman or VX were survived by ~20% of the animals. In the 2 and 4 week exposure study with 0.2 and 0.43LD50 of each agent, blood AChE was inhibited from 75% to 91% of control activity on the last day of exposure depending upon the agent and the dose regardless of whether the animals were exposed for 2–4 weeks. Diaphragm AChE measured 3 days after the last exposure was inhibited 18%–30% depending upon the agent, but the differences were not statistically significant compared with controls. There were no differences in body weight gains, body temperature, blood chemistries, hematology, or histopathology of any of the tissues collected between exposed guinea pigs and the controls. It should be noted that in the 4 week exposure study, animals received cumulative doses of 4 and 83the acute LD50 of each of these nerve agents.

The 2 week repeated exposure paradigm in guinea pigs has been used in subsequent studies of the effects of repeated nerve agent exposure on learned and unlearned behaviors, EEG activity, and brain neurochemistry. Hulet and colleagues (Hulet et al., 1993; Shih et al., 2006)[76-77] used this exposure protocol with guinea pigs to determine any adverse effects of repeated doses (saline, 0.43or 0.53LD50) of sarin. The animals were assessed for changes in body weight, RBC–AChE levels, neurobehavioral reactions to an FOB, cortical EEG power spectrum, and intrinsic ACh neurotransmitter regulation over the 2 weeks of sarin exposure and for up to 12 days postinjection. Animals receiving the 0.53LD50 showed signs of cholinergic toxicity and gained significantly less weight than the 0.43LD50 or saline animals, which did not differ from each other. RBC–AChE levels had dropped to 11% and 10% of control values for the 0.4 and 0.53LD50 groups, respectively, by the end of the sarin-exposure period, although the rate of AChE depression was faster in the 0.53LD50 group. Both the 0.4 and 0.53LD50 groups failed to display evidence of habituation to aspects of being handled over the period of exposure, unlike the controls which showed a significant decline in these scores on the FOB test. Power spectral analysis of the EEG showed that saline-injected animals tended to fall asleep after exposure and showed a progressive increase in the power of the (delta, theta) bandwidths. The high-dose sarin-exposed animals (0.53LD50) failed to show any change in distribution in EEG power over this time while the 0.43LD50 sarin animals were intermediate between the controls and the high-dose group. These changes persisted for 6 days after the exposure ended for the high-dose group. Six days after the end of the exposure, sarin-exposed animals showed a greater increase in striatal ACh output than controls, as measured by microdialysis, in response to a pharmacological challenge with atropine. The authors concluded that the results suggested subtle, but transient, effects on neurobehavioral function, EEG, and brain neurochemistry for the asymptomatic animals that received the 0.43LD50 sarin, and possibly more persistent changes in the symptomatic 0.53LD50 sarin group.

Langston et al. (2005) [78] trained dietary restricted guinea pigs responding for food on progressive ratio (PR) schedule. Once performance had stabilized, the animals were exposed daily (5 day=week) to 0.1, 0.2, or 0.43LD50 sarin for 2 weeks and performance was measured during the 2 weeks of exposure and for 2 weeks following the exposure. The two lower doses (0.1 and 0.23LD50) produced no effects on performance during or after the period of exposure. However, the 0.43LD50 dose group showed decreases in body weight and concurrent decreases in response rates and break points that was due to increases in pausing following receipt of food reinforcement. These performance changes reverted to preexposure levels following termination of the exposure. There are several other reports that have demonstrated the disruptive effects of the 0.43LD50 chronic sarin, soman, or VX exposure in food-restricted animals, which has also been reported by others. Sipos et al. (2001; and in Thompson et al., 2004, 2005)[79-81] evaluated the effects of repeated lowdose exposure of guinea pigs to sarin, soman, or VX using this same procedure and evaluate acoustic startle response and active avoidance (two-way; air puff) performance. Chronic low-level exposure to sarin, soman, or VX enhanced acoustic startle.

# CHAPRTER 8 DIAGNOSIS AND DETECTION OF NERVE GAS AGENTS

## 8.1 Diagnosis of Nerve gas poisoning

The diagnosis of nerve agent exposure is made when there is a high index of suspicion and when the symptoms described above present themselves. Evaluation for nerve agent poisoning should occur only after rinsing the victim completely with water. A dermal or respiratory exposure route may dictate a different constellation of symptoms and time course of events, depending upon the amount of the exposure. Exposure of the skin to liquid agent may be initially asymptomatic in the first 10-30 minutes. This is then followed quickly by respiratory and neurologic effects. If the amount is small the onset of localized symptoms such as sweating and fasciculations not occur until 18 hours later. Therefore, patients must be observed for this length of time when the suspicion is high that there has been an exposure to reach a diagnosis. Respiratory symptoms on the other hand occur immediately with inhalation of nerve agent. Large amounts will cause respiratory failure within seconds to minutes whereas small amounts cause only limited ocular and airway effects such as miosis, eye pain, bronchorrhea, and bronchospasm.

Blood should be drawn for red blood cell AChE activity (nerve agents inactive cell activity over that of the plasma activity). If this red blood cell AChE is reduced by 70%, there is a 50% likelihood of systemic effects.

## 8.2 Detection of nerve gas poisoning

Nerve agents may be detected by a variety of means. Single and three colour detector papers are available for individual issue to detect liquid nerve agent. Area detectors and monitors of various sensitivities are available.

The following detectors are the most commonly used for detection of nerve gas.

a) Lightweight Chemical Agent Detector (LCAD)

b) Man-Portable Chemical Agent Detector

c) Nerve Agent Immobilised-Enzyme and Detector (NAIAD)

**Figure of the detector**

**Figure10 Lightweight Chemical Agent Detector (LCAD)**

**Figure11 Man-Portable Chemical Agent Detector**

**Figure12 Nerve Agent Immobilised-Enzyme and Detector (NAIAD)**

# CHAPTER 9 TREATMENT AND MANAGEMENT OF NERVE GAS POISONING

The medical management of nerve agent poisoning consists of pre-treatment, diagnosis and treatment with a combination of pharmacological, respiratory and other general supportive measures, as dictated by the condition of the casualty. The life-threatening effects of acute nerve agent poisoning occur much more rapidly and for a generally shorter duration than those produced by organophosphate (OP) insecticides. Following pre-treatment, successful post-exposure treatment relies on equally timely administration of necessary therapeutic measures. Urgent skin decontamination and continued respiratory protection must be applied alongside the administration of autoinjectors for self and buddy aid, all of which must be maintained during rapid evacuation to medical care.

Triage priorities must be reviewed frequently, with subsequent delivery of appropriate drug therapy and continue respiratory support, where necessary. This must be administered in a timely and confident manner if emergency medical management is to be successful and the ultimate prognosis favourable. Field medical management may have to be conducted initially in a contaminated environment, requiring practised procedures to avoid cross-contamination and unnecessary exposure of the already compromise casualty.

## 9.1 Pre-treatment

Poisoning by nerve agents which form rapidly ageing complexes may be particularlydifficult to treat. Ageing is the biochemical process by which the agent-enzyme complex becomes completely resistant to oxime reactivation. In the case of soman (GD), the agent-enzyme complex ages within minutes. These difficulties have been mitigated, in part, by the use of carbamates as pretreatment.

The terms pre-treatment or prophylaxis are differentiated thus:- Pre-treatment: the administration of drugsin advance of poisoning designed to increase the efficacy of treatment administered post-poisoning.

- Prophylaxis: the administration of drugs in advance of the poisoning designed to make post-poisoning therapy unnecessary.

Current pretreatments utilise carbamate anticholinesterases, for example, pyridostigmine, which may be used either alone or in combination with other anticholinergic drugs. They bind cholinesterases *reversibly*,preventing the organophosphate binding to the enzyme. The term *reversible* is used here comparatively: the carbamateacetylcholinesterase complex breaks down fairly rapidly, while organophosphateacetylcholinesterase complexes break down very slowly. In comparison, the aged organophosphate acetylcholinesterase complex is stable.

When carbamates are used as pretreatments, carbamoylation of acetylcholinesterase prevents phosphorylation, butlater the carbamate-acetylcholinesterase complex dissociates, freeing active enzyme. Current pre-treatment regimen (pyrdistigmine,30 mg 8 hourly) are designed to give 20 – 40% binding of peripheralacetylcholinesterase, thereby allowing the carbamate to protect a proportion of the agent. There are no significant signs and there are no significant signs and symptoms associated with this level of inhibition since

- the process of inhibition of the enzyme is slower than that due to nerve agent.

- there is negligible inhibition of centralacetylcholinesterase.

In conjunction with post exposure therapy, good protection against lethality is obtained within 2 hours of the first dose, but is enhanced by continued intake of pyridostigmine pre-treatment, according to the accepted dosage regimen.Pyridostigmine pre-treatment should be stopped when the symptoms of nerve agent poisoning following a chemical warfare attack are observed; post exposure therapy must be started immediately.

Pyridostigmine is not an antidote, and it should not be taken after nerve agent exposure. Its continued use post-exposure may well enhance the effects of nerve agent poisoning. It is ineffective unless standard therapy is also used in the appropriate manner.

The recognised side effects of prolonged use of pyridostigmine include gastrointestinal intestinal changes including increased flatus, loose stools, abdominal cramps and nausea. Other reported effects are urinary urgency, headache, rhinorrhoea, diaphoresis and tingling of the extremities.

However, several studies have concluded that these side effects are tolerable by most personnel and lead to negligible degradation of military performance - symptoms due to pyridostigmine may be ameliorated by taking the tablets with food, and Pyridostigmine pre-treatment was discontinued on medical advice in less than 0.1% of individuals, generally because of intolerable nausea and diarrhoea.

When taken in excess of the recommended dosage, symptoms of carbamate poisoning will occur. These include diarrhoea, gastrointestinal cramps, tight chest, nausea, rhinorrhoea, headache and miosis. Good compliance is required if optimal protection is to be obtained. The importance of Pyridostigmine pre-treatment should therefore be stressed during training.

Medical officers must be aware of potential drug interactions and contraindications when dealing with personnel on pretreatment. This is of particular relevance with respect to some muscle relaxants used in anaesthesia. Pyridostigmine has been used in clinical situations for up to 50 years. At this time there is no evidence of long term side effects.

## **9.2 Post-Exposure Therapy**

The main principles of post exposure treatment for nerve agent poisoning are based on early drug therapy in combination where necessary with measures to support respiration and provide general supportive care.

## 9.3 Pharmacological Treatment of Nerve Agent Poisoning

The pharmacological treatment of nerve agent poisoning involves the use of:

- Anticholinergics to antagonise the muscarinic effects.

- Oximes to reactivate inhibited enzyme and antagonise the nicotinic effects.

- Anticonvulsants to prevent seizure activity and protect against subsequent CNS damage.

The effects of drugs used in nerve agent poisoning are described below.

### 9.3.1 Anticholinergic – Atropine

Atropine sulphate is an essential drug in the treatment of nerve agent poisoning. It acts by blocking the effects of acetylcholine at muscarinic receptors and so produces relief from many of the symptoms previously listed. Some therapeutic effects are also produced within the central nervous system; at high doses (> 6 mg) atropine may help reduce central respiratory depression, and if administered early after poisoning (5 – 10 minutes), may have some beneficial anticonvulsant action.

#### 9.3.1.1 Dose Regimen

Immediate treatment with atropine in cases of systemic nerve agent poisoning is essential, and the absolute requirement will depend on both the severity of effects and whether pre-treatment has been used.

Following autoinjector therapy for first aid doses should be repeated regularly until signs of successful atropinisation are noted. Intervals of 5 – 15 minutes may be required for continuing but not life-threatening effects of poisoning; the intravenous route of administration is preferred. In cases of severe poisoning, higher doses may well be required. Atropine is given intravenously at a rate of 2 mg every 3-5 minutes until the casualty is atropinised. Subcutaneous or intramuscular routes of administration are alternatives, and recommended until significant hypoxia is corrected, as in such cases intravenous administration of atropine may precipitate life-threatening ventricular arrythmias. In the Iran-Iraq War, intravenous infusion of atropine at 2 mg/min for over an hour was administered with considerable success for casualties severely poisoned by nerve agent vapour who had not received pre-treatment.

In severe cases that have previously received pre-treatment, it is estimated that 30 – 50 mg of atropine in total may be required to achieve atropinisation.

Signs of successful atropinisation include decreased bronchospasm, reduced airways resistance, the drying-up of bronchial and salivary secretions, reduced sweating and a stabilisation in the heart rate to around 90 beats per minute. The effect of atropine in drying bronchial secretions may make the removal of mucus more difficult, so suction is likely to be necessary.

After the emergency field treatment, atropinisation should be maintained for at least 24 hours by intramuscular injection or slow intravenous infusion of 1 to 2 mg of atropine per hour as required.

### 9.3.1.2 Complications

Atropine, especially in the presence of hypoxia, may render the myocardium more susceptible to arrhythmias. Correction of hypoxia, and ECG monitoring if available, is essential during atropinisation. Overdosage may produce euphoria, hallucinations, anxiety, and delirium. Close observation is necessary and sedation of casualties may be required. Bladder dysfunction may necessitate catheterisation. By the reduction of sweating, atropine increases the risk of heat stress.

### 9.3.1.3 Inappropriate Administration

If atropine is administered in the absence of nerve agent poisoning, the following effects may be noted:

- dryness of the mouth and pharynx
- decreased sweating
- slight flushing and tachycardia

- some hesitancy of micturition

- slightly dilated pupils

- mild drowsiness

- impaired memory and recal

- blurring of near vision

After 2 mg, these symptoms should not interfere with ordinary activity except in the occasional person, in hot environments or at high work rates. Higher doses, or repeated doses, will produce more marked symptoms which may become totally incapacitating, particularly in warm environments or high work rates. The effects of atropine are fairly prolonged, lasting 3 to 5 hours after one or two injections of 2 mg and 12 to 24 hours after significant over-atropinisation.

The traditional antidote for atropine overdose is physostigmine. This is a carbamate anticholinesterase, like pyridostigmine, but crosses the blood brain barrier.

As physostigmine acts by inhibiting anticholinesterase, it must be used with caution in treating over-atropinisation following nerve agent poisoning, where the levels of enzyme may be reduced below normal. In such cases, the risk of repeated cholinergic effects is greater. Its use should therefore be reserved only for those cases with severe symptoms of atropine overdose, refractive to general supportive measures.

### 9.3.1.4 Atropine For Eye Effects

Atropine given parenterally has comparatively little effect on nerve agent induced miosis. The local application of cycloplegics (homatropine or similar eye drops) to the eye reduces both the degree of miosis, eye pain and associated nausea, vomiting and headache. However, expert opinion on the value of atropine-containing eye drops in the management of nerve agent induced miosis remains divided. It is believed by some that problems of accommodation may be made worse by the application of the drops and that, overall, little benefit may be produced.

### 9.3.1.5 Other Anticholinergics

On the basis of experimental data, other anticholinergic drugs are considered to be beneficial in the treatment of nerve agent poisoning. Anticholinergics with pronounced central antimuscarinic effects (e.g. benactyzine, biperiden, trihexyphenidyl, scopolamine) have the potential to suppress seizure activity without further anticonvulsant therapy, if administered early after poisoning.

Drug combinations with atropine (e.g. benactyzine and atropine) have been introduced by some nations in specific clinical situations, but caution must be exercised when using these regimen.

## 9.4 Oximes

### 9.4.1 Toxicity

While atropine blocks the cholinergic effects of nerve agent poisoning at muscarinic sites, it has little effect upon the nicotinic effects at skeletal neuromuscular junctions and the autonomic ganglia. Amelioration of nerve agent effects at these sites and also at muscarinic sites can, however, obtained by the reactivation of inhibited AChE by means of oximes.

Three oximes are available in autoinjectors for self- or buddy-administration and these are:

- N-methyl derivative of 2-pyridine-2- aldoxime (pralidoxime) as the methanesulphonate; methyl sulphate and chloride salt.

- toxogonin (obidoxime) dichloride.

- HI-6 chloride.

Note that atropine and oximes are synergistic in their effects and the administration of atropine and oxime in combination has been shown experimentally to raise the LD50's of some nerve agents by a factor in excess of 20.

## 9.5 Enzyme Reactivation

The relative potency of the different oximes in reactivating inhibited AChE varies according to the nature of the nerve agent.

The choice of oxime (if more than one is available) is not only dependent on the identity of the nerve agent responsible for the poisoning; the effectiveness of a given oxime is also dependent on whether the casualty has received pyridostigmine pre-treatment and the administration of other anticholinergic and anticonvulsant drugs.

In the field, immediate post-exposure therapy consisting of atropine, oxime and diazepam will be given by intramuscular injection from an autoinjector device on the appearance of the first significant signs of nerve agent poisoning. Following the use of the autoinjector, if casualties remain symptomatic, they will require further doses of oxime (and atropine) to alleviate the clinical situation and reactivate inhibited enzyme. Opinion remains divided as to the most appropriate way of achieving therapeutic blood levels of oxime.

The intravenous injection of bolus doses of oxime is probably the only practical solution on the battlefield.

Under field conditions, single bolus doses can also be given by intramuscular injection.

An alternative method of administering therapeutic doses of oxime is by intravenous infusion. Sundwall reported experimentally that a plasma concentration of 4 $\mu g.ml^{-1}$ pralidoxime methanesulphonate was needed to counteract the neuromuscular block, bradycardia, hypotension and respiratory failure caused by organophosphate anticholinesterases. This concentration has been assumed since then to be the minimum concentration of oxime required to counteract nerve agent intoxication in man. Based on this assumption, it is possible to calculate the loading and maintenance doses for the intravenous administration of oximes that will achieve this level.

There is only limited experience in human poisoning by organophosphorus nerve agents but it is generally accepted that the persistence of clinically relevant amounts of nerve agent in the blood is shorter than that of insecticides or pesticides. However, as a result of ageing of the inhibited AchE (which may be very rapid following poisoning with GD), it is suggested that in the absence of clinical improvement, administration of oxime for periods in excess of 24-48 hours is unlikely to achieve further reactivation of the enzyme.

## 9.6 Anticonvulsants

In severe poisoning by nerve agent, atropine protects only partially against seizure activity and the resulting brain damage; other anticholinergics vary in this ability. Early administration of anticonvulsants should be mandatory.

Experimental evidence shows that the early administration (5 – 10 min) of the benzodiazepines antagonises the seizure activity of nerve agent. The addition of this class of drugs to the basic treatment regimen greatly improves morbidity and mortality, in addition to its anti-seizure effect. Currently Diazepam is the most widely fielded and should be administered as a 5 - 10 mg dose initially and further doses should be given frequently enough to control seizures. This may require additional injections, at intervals ranging from a few minutes to several hours.

In the absence of anticonvulsant therapy, irreversible brain damage may result; this is exacerbated by periods of hypoxia.

## 9.7 Respiratory Supportive Care

### 9.7.1 Airway management

A patent airway should always be maintained, with postural drainage and suction. In an unconscious casualty, adequate atropine should be administered to reduce airway resistance. Cricothyroidotomy and/or endotracheal intubation may be needed.

## 9.7.2 Respiratory support

Dyspnoeic casualties may require respiratory support; this should be started at the lowest possible level. In a casualty that is dyspnoeic, the respirator should be removed and replaced with a casualty hood which will reduce the ventilatory resistance.

## 9.7.3. Positive pressure ventilation

Assisted ventilation may range from mouth to- mouth ventilation (in the absence of a liquid or vapour hazard and after appropriate decontamination) to the use of an automatic ventilator.

### 9.7.3.1 Oxygen

Dyspnoeic casualties should receive oxygen at the lowest possible level. This may result in a high demand for oxygen. If available, oxygen saturation should be monitored.

### 9.7.3.2 General Supportive Care

Although pre- and post-exposure therapy will reduce lethality, casualties may still be incapacitated. A patient severely poisoned by an anticholinesterase is a critical medical emergency and may require intensive care for days or weeks. General supportive care such as restoring fluid and electrolyte balance, correction of acid base balance, control of body temperature and control of infection should follow normal medical practice.

Special precautions should be taken when using muscle relaxants in patients poisoned by nerve agents.

## 9.8 Emergency Field Therapy

### 9.8.1 Self Aid (or Buddy Aid)

This comprises first aid measures which the soldier can apply to help him- or herself. The rapid action of nerve agents calls for immediate self treatment. Unexplained nasal secretion, salivation, tightness of the chest, shortness of breath, constriction of pupils, muscular twitching, or nausea and abdominal cramps call for the immediate intramuscular injection of 2 mg of atropine, combined if possible with oxime. From 1 to 3 automatic injection devices, each containing 2 mg atropine or mixture of atropine, oxime and/or anticonvulsant, are carried by each individual.

One device should be administered immediately when the symptoms and/or signs of nerve agent poisoning appear. This may be done by the casualty or by a buddy, the injection being given perpendicularly through the clothing into the lateral aspect of the middle of the thigh. Further devices, up to a total of 3, should be administered by the casualty or by his or her buddy during the following 30 minutes if the symptoms and/or signs of poisoning fail to resolve.

The timing of these further injections and whether they are given at one time or separately may depend on the casualty's condition (and on instructions promulgated by individual NATO nations).

Note that if automatic injectors are used in the absence of exposure to agent, the following signs and symptoms may be seen: dry mouth, dry skin, fast pulse (>90 beats per minute), dilated pupils, retention of urine and central nervous system disturbance.

Susceptibility to heat exhaustion or heat stroke is increased, particularly in closed spaces or whilst wearing protective clothing.

### 9.8.2 First Aid by Trained Personnel

This comprises the emergency actions undertaken to restore or maintain vital bodily functions in a casualty. Wherever the casualty is not masked, the respirator must be adjusted for him or her by the nearest available person. Attention should be given to decontamination at the earliest possible moment and any skin contamination must be removed with a personal decontamination kit.

After nerve agent poisoning, the administration of atropine is repeated at intervals until signs of atropinization (dry mouth and skin and tachycardia >90 per minute) are achieved. Miosis from vapour exposure is not relieved by systemic atropine.

Mild atropinization should be maintained for at least 24 hours by intramuscular injection of 1-2 mg of atropine at intervals of 1/2 to 4 hours, as required. There is a danger of ventricular arrhythmias arising from atropinization if the casualty is anoxic.

Assisted ventilation is required for severely poisoned individuals as they will have:
- Marked bronchoconstriction.
- Copious secretions in the trachea and bronchi.
- Paralysis of the respiratory muscles.
- Central respiratory depression, hypoxia, and convulsions.

### 9.8.3 Assisted Ventilation

Positive pressure resuscitation should be undertaken but the pressure necessary to overcome the bronchoconstriction may be more than 65 cm H2O and if possible, intubation is highly desirable. In the absence of a liquid or vapour hazard, assisted ventilation may be attempted by the standard mouth-to-mouth method after decontamination of the casualty's face and mouth.

In the presence of a liquid or vapour hazard, ventilation may be given by a portable resuscitator with NBC filter attached.

In a medical care facility, mechanical resuscitation of the positive pressure type may be used with endotracheal intubation or tracheostomy; artificial respiration must be continued until the casualty is breathing normally or the medical personnel have pronounced the casualty dead. Due to the production of copious secretions, regular suction will be required.

# CHAPTER 10 PROGNOSIS OF NERVE GAS EXPOSURE

Recovery from mild exposures is rapid; however, 10 % may have continued symptoms up to 3 weeks or beyond. Should the patient with symptoms survive the immediate effects of nerve agent poisoning, several delayed effects may occur which need to be identified should they occur. There is a delayed intermediate syndrome which can occur a few days following exposure and immediate recovery that may require re-initiation of respiratory support measures including intubation. This manifests with onset of descending muscular paralysis.

Another delayed reaction is organophosphate induced delayed neuropathy (OPIND). This is due to slow inhibition of neuropathy target esterase (NTE) which can occur up to a month after exposure. The function of NTE is not known. It is thought that inhibition of NTE leads to a sensory and motor distal neuropathy that is similar to Guillian-Barré syndrome.

# CHAPTER 11 INFAMOUS SARIN ATTACK ON THE TOKYO SUBWAY

In five coordinated attacks, the perpetrators released sarin on several lines of the Tokyo Metro, killing thirteen people, severely injuring fifty and causing temporary vision problems for nearly a thousand others. The attack was directed against trains passing through Kasumigaseki and Nagatachō, home to the Japanese government.

## BACKGROUND

Aum Shinrikyo is the former name of a controversial group now known as Aleph. In 1992 Shoko Asahara, the founder of Aum Shinrikyo published a landmark book, in which he declared himself "Christ", Japan's only fully enlightened master and identified with the "Lamb of God". He outlined a doomsday prophecy, which included a Third World War, and described a final conflict culminating in a nuclear" Armageddon", borrowing the term from the Book of Revelation 16:16. His purported mission was to take upon himself the sins of the world, and he claimed he could transfer to his followers spiritual power and ultimately take away their sins and bad work. He also saw dark conspiracies everywhere promulgated by Jews, Freemasons, the Dutch, the British Royal Family, and rival Japanese religions. The Japanese police initially reported that the attack was the cult's way of hastening an apocalypse. The prosecution said that it was an attempt to bring down the government and install Shoko Asahara, the group's founder, as the "emperor" of Japan. Asahara's defense team claimed that certain senior members of the group independently planned the attack, but their motives for this were left unexplained. Aum Shinrikyo first began their attacks on 27 June 1994 in Matsumoto, Japan. With the help of a converted refrigerator truck, members of the cult released a cloud of sarin which floated near the homes of judges who were overseeing a lawsuit concerning a real-estate dispute which was predicted to go against the cult. From this one event, 500 people were injured and seven people died

### Shoko Asahara

## Attack

On Monday March 20, 1995, five members of Aum Shinrikyo launched a chemical attack on the Tokyo Metro, one of the world's busiest commuter transport systems, at the peak of the morning rush hour. The chemical agent used, liquid sarin, was contained in plastic bags which each team then wrapped in newspaper. Each perpetrator carried two packets totaling approximately 900 millilitres of sarin, except Yasuo Hayashi, who carried three bags. Aum originally planned to spread the sarin as an aerosol but did not follow through with it. A single drop of sarin the size of a pinhead can kill an adult.

Carrying their packets of sarin and umbrellas with sharpened tips, the perpetrators boarded their appointed trains. At prearranged stations, the sarin packets were dropped and punctured several times with the sharpened tip of the umbrellas. Each man then got off the train and exited the station to meet his accomplice with a car. By leaving the punctured packets on the floor, the sarin was allowed to leak out into the train car and stations. This sarin affected passengers, subway workers, and those who came into contact with them. Sarin is the most volatile of the nerve agents, which means that it can quickly and easily evaporate from a liquid into a vapor and spread into the environment. People can be exposed to the vapor even if they do not come in contact with the liquid form of sarin. Because it evaporates so quickly, sarin presents an immediate but short-lived threat.

On the day of the attack, ambulances transported 688 patients and nearly five thousand people reached hospitals by other means. Hospitals saw 5,510 patients, seventeen of whom were deemed critical, thirty-seven severe and 984 moderately ill with vision problems. Most of those reporting to hospitals were the "worried well", who had to be distinguished from those who were ill.

By mid-afternoon, the mildly affected victims had recovered from vision problems and were released from hospital. Most of the remaining patients were well enough to go home the following day, and within a week only a few critical patients remained in hospital. The death toll on the day of the attack was eight.

This accident was an eye opener regarding the precautions and measures that have to be taken for incidents like this.

# CHAPTER 12 FUTURE PROSPECTS FOR IMPROVEMENT REGARDING NERVE GAS ATTACK AND PREVENTATION.

In their paper "Emergency response planning for a Potential Sarin Gas attack in Manhanttan using agent based models" *Venkatesh Mysore et al* described the agent based modeling (ABM),simulation and analysis of a potential Sarin gas attack in the Port Authority Bus Terminal in the island of Manhanttan in the New York city, USA. The streets and subways of Manhattan had been modeled as a non planar graph. The people at the terminal were modeled as agents initially moving randomly, but with a resultant drift velocity towards their destinations e.g. work places. Upon exposures and illness, they choose to head to one of the hospitals they are aware of. Authors given a simple variant of the LRTA algorithm for route computation were used to model a person's panic behavior.

Authors had described various mathematical calculations which describes what are the factors that were to taken in consideration while designing a master plan for attack like this.

The current drugs which are used for treatment of poisoning nerve gas suffers from few drawbacks like the adverse reactions caused by the drug like elevated heart rate, blur vision, dizziness, fatigue etc. Researcher should find alternative for these drugs in herbal and traditional system of medicine so that the treatment of sarin gas could be more efficacious and having less adverse reactions.

## **CHAPTER 13 CONCLUSION**

Chemical warfare using nerve gas is one of the most dangerous problem encountered by the modern society. The use of nerve agents should be abolished else one day this whole world should be convert into a desert of flesh and blood.

## **References**

1) Tagate. (2006). Available athttp:==www.tagate.com=wars=history_of_warfare=chemical_warfare_4.shtml (accessed January 2007).

2) Wikipedia. (2007b). Greek fire. Available from URL: http:==www.wikipedia.org=wiki=Greek_fire (accessed January 2007).

3) Geiling, J.A. (2003). Chemical Terrorism. Available at http:==www.chestnet.org=education=online=pccu= vol16=lessons17_18=lesson18=print.php (accessed January 2007).

4) DeNoon, D. (2004). Biological and chemical terror history. Available at URL: http:==www.webmd.com=content= article=61=67268.htm?printing¼true (accessed January 2007).

5) Mauroni, A. (2003). Chemical and Biological Warfare. ABC CLIO Publishers. p. 80.

6) Hemsley, J. (1987). The Soviet Biochemical Threat to NATO. St. Martin's Press, New York. pp. 14, 66–67.

7) United Kingdom Ministry of Defense. (1999). Defending against the threat of biological and chemical weapons. Outline history of biological and chemical weapons.

8) Noji, E.K. (2001). Bioterrorism: a 'new' global environmental health threat. Global Change and Human Health 2(1): 46–53.

9) Robey, J. (2003). Bioterror through time. Discovery Channel Series, February 21, 2003.

10) Tschanz, D.W. (2003). ''Eye of newt and toe of frog'': biotoxins in warfare. StrategyPage.com, October 20, 2003.

11) Thucydides. (1989). The Peloponnesian War. D. Grene, Ed. The University of Chicago Press, Chicago. pp. 134, 286–287.

12) Hickman, D.C. (1999). Seeking Asymmetric Advantage: Is Drinking Water an Air Force Achilles' Heel? Air Command and Staff College, Air University, Maxwell Air Force Base, Alabama. Research Report AU=ACSC=084=1999–04.

13) CNS. (2001). Chronology of State use and biological and chemical weapons control. Available from http:==www.cns.miis.edu=research=cbw=pastuse.htm (accessed January 2007).

14) Miles, W.D. (1957b). The chemical shells of Lyon Playfair. Armed Forces Chemical Journal 11(6): 23.

15) Camerman, N. and Trotter, J. (1963). Stereochemistry of arsenic: VIII. Canadian Journal of Chemistry 41: 460–464.

16) Sartori, M. (1939). The War Gases: Chemistry and Analysis. D. Van Nostrand Co. Inc. NY. pp. 3, 33, 59, 165, 181, 188, 217.

17) Field Manual 3–11.9. (2005). Potential Military Chemical=Biological Agents and Compounds. I-2, II-9, 15, 18, 21, 31–33, 37–38, 50, III-13.

18) Rozman, K.K. and Doull, J. (2001). Paracelsus, Haber and Arndt. Toxicology. 160: 191–196.

19) Charles, D. (2005). Master Mind. Harper Collins Publishers. pp. 152, 154–157.

20) SIPRI. (1971). The Rise of CB Weapons. Humanities Press, New York, 1971, pp. 70–75, 142, 147, 280–282, 336–341.

21) Harris, R. and Paxman, J. (1982). A Higher Form of Killing: The Secret Story of Chemical and BiologicalWarfare. New York: Hill and Wang, New York. pp. 24, 117–123, 141.

22) Pringle, L. (1993). Chemical and Biological Warfare. Enslow Publishers, Inc., Hillside, NJ. p. 55, 57.

23) Yang, Y.-C., Baker, J.A., and Ward, J.R. 1992. Decontamination of chemical warfare agents. Chemical Reviews, 92: 1729–1743.

24) Wikipedia

25) Henderson, T.J. 2002. Quantitative NMR spectroscopy using coaxial inserts containing a reference standard: purity determinations for military nerve agents. Analytical Chemistry, 74: 191–196.

26) Henderson, T.J. 2002. Quantitative NMR spectroscopy using coaxial inserts containing a reference standard: purity determinations for military nerve agents. Analytical Chemistry, 74: 191–196. 27) Wagner, G.W., Bartram, P.W., Koper, O., and Klabunde, K.J. 1999. Reactions of VX, GD and HD with nanosize MgO. Journal of Physical Chemistry B, 103: 3225–3228.

27) Katritzky, A.R., Duell, B.L., Durst, H.D., and Knier, B.L. 1988. Substituted o-iodoso- and o-iodoxybenzoic acids: synthesis and catalytic activity in the hydrolysis of active phosphorus esters and related systems. Journal of Organic Chemistry, 53: 3972–3978.

28) Ketelaar, J.A.A., Gersmann, H.R., and Beck, M.M. 1956. Metal-catalyzed hydrolysis of thiophosphoric esters. Nature, 177: 392–393.

29) Albizo, J.M. and Ward, J.R. 1988. Hydrolysis of GD and VX by 0.05M=0.10M copper(II) N,N,N0,N0-tetramethylenediamine (TMEN), in Proceedings, Army Science Conference (16th), Vol. 1, 25–27 October 1988: 33–37, AD-A203101.

30) Yang, Y. 1999. Chemical detoxification of nerve agent VX. Accounts of Chemical Research, 32: 109–115.

31) Davisson, M.L., Love, A.H., Vance, A., and Reynolds, J.G. 2005. Environmental Fate of OrganophosphorusCompounds Related to Chemical Weapons. UCRL-TR-209748.

32) Yesodharan, S. 2002. Supercritical water oxidation: an environmentally safe method for the disposal of organic wastes. Current Science, 82: 1112–1122.

33) Borkin, J. 1997. The Crime and Punishment of I.G. Farben. New York: Barnes and Noble.

34) Kosolapoff, G.M. 1950. Organophosphorus Compounds, New York: John Wiley.

35) Benes, J. 1963. Kinetics of the hydrolysis of tabun in acid solutions. Acta Chemica Scandinavica, 17:1783–1785.

36) Sanchez, M.L., Russell, C.R., and Randolph, C.L. 1993. Chemical Weapons Convention Signature Analysis, DNA-TR 92–73ADB171788. Alexandria, VA: Defense Technical Information Center.

37) McNaughton, M.G. and Brewer, J.H. 1994. Environmental Chemistry and Fate of Chemical Warfare Agents.SWRI Project 01–5864. San Antonio, TX: Southwest Research Institute.

38) D'Agostino, P.A.D. and Provost, L.R. 1992. Mass spectrometric identification of products formed during degradation of ethyl dimethylphosphoramidocyanidate (tabun). Journal of Chromatography, 598: 89–95.

39) Larsson, L. 1952. A spectrophotometric study in infra-red of the hydrolysis of dimethylamido-ethoxy-phosphoryl cyanide (tabun). Acta Chemica Scandinavica, 6: 1470–1476; and references cited therein;, 1953, 7: 306–314.

40) Degenhardt, C.E.A.M., Van Den Berg, G.R., De Jong, L.P.A., Benschop, H.P., Genderen, J.V., and Van de Meent, D. 1986. Enantiospecific complexation gas chromatography of nerve agents: isolation and properties of the enantiomers of ethyl N,N 0-dimethylphosphoramidocyanidate (tabun). Journal of the American Chemical Society, 108: 8290–8291.

41) Van Den Berg, G.R., Beck, H.C., and Benschop, H.P. 1984. Stereochemical analysis of the nerve agents soman, sarin, tabun, and VX by proton NMR-spectroscopy with optically active shift reagents. Environmental Contamination Toxicology, 33: 505–514.

42) Albaret, C., Loeillet, D., Auge, P., and Fortier, P.-L. 1997. Application of two-dimensional 1H-31P inverse NMR spectroscopy to the detection of trace amounts of organophosphorus compounds related to the chemical weapons convention. Analytical Chemistry, 69: 2694–2700.

43) Sidell, F.R., Patrick, W.C., and Dashiell, T.R. 1998. Jane's Chem-Bio Handbook. Alexandria, VA: Jane's Information Group: 92.

44) Black, R.M. and Harrison, J.M. 1996a. The chemistry of organophosphorus chemical warfare agents, in The Chemistry of Organophosphorus Compounds, Vol. 4, edited by F.R. Hartley, New York: John Wiley and Sons: 791.

45) Bryant, P.J.R., Ford-Moore, A.H., Perry, B.J., Wardrop, A.W.H., and Watkins, T.F. 1960. The preparation and physical properties of isopropyl methylphosphonofluoridate (Sarin). Journal of the Chemical Society, 1960: 1553–1555.

46) Reesor, J.B., Perry, B.J., and Sherlock, E. 1960. The synthesis of highly radioactive methylphosphonofluoridate (Sarin) containing P32 as a tracer element. Canadian Journal of Chemistry, 38: 1416–1427.

47) Boter, H.L., Ooms, A.J.J., Van Den Berg, G.R., and Van Dijk, C. 1966. The synthesis of optically active isopropyl methylphosphonofluoridate (Sarin). Recueil des travaux chimiques des Pays-Bas, 85: 147–150.

48) Epstein, J., Bauer, B.V., Saxe, M., and Demek, M.M. 1956. The chlorine-catalyzed hydrolysis of isopropyl methylphosphonofluoridater (Sarin) in aqueous solution. Journal of the American Chemical Society, 78: 4068–4071.

49) Dubey, D.K., Gupta, A.K., Sharma, M., Prabha, S., and Vaidyanathaswamy, R. 2002. Micellar effects on hypochlorite catalyzed decontamination of toxic phosphorus esters. Langmuir, 18: 10489–10492.

50) Demek, M.M., et al. 1970. Behavior of Chemical Agents in Seawater, EATR 4417. August 1970, AD-873242.

51) Bard, J.R., Daasch, L.W., and Klapper, H. 1970. Kinetics of reaction between isopropyl methylphosphonofluoridate and hydrogen chloride. Journal of Chemical and Engineering Data, 15: 134–137.

52) Larsson, L. 1958b. A kinetic study of the reaction of isopropoxy-methyl-phosphoryl fluoride (Sarin) with hydrogen peroxide. Acta Chemica Scandinavia, 12: 723–730.

53) Christen, P.J. and Van Den Muysenberg, J.A. 1965. Enzymic isolation and fluoride catalyzed racemization of optically active Sarin. Biochimica Biophyica Acta, 210: 217–220.

54) Yang, Y.-C., Baker, J.A., and Ward, J.R. 1992. Decontamination of chemical warfare agents. Chemical Reviews, 92: 1729–1743.

55) Moss, R.A., Alwis, K.W., and Bizzigotti, G.O. 1983. o-Iodosobenzoate: catalyst for the micellar cleavage of activated esters and phosphates. Journal of the American Chemical Society, 105: 681–682.

56) Moss, R.S., Alwis, K.W., and Shin, J.-S. 1984. Catalytic cleavage of active phosphate and ester substrates by iodoso- and iodoxybenzoates. Journal of the American Chemical Society, 106: 2651–2655.

57) Moss, R.A., Kim, K.Y., and Swarup, S. 1986. Efficient catalytic cleavage of reactive phosphates by an o-iodosobenzoate functionalized surfactant. Journal of the American Chemical Society, 108: 788–793.

58) Katritzky, A.R., Duell, B.L., Durst, H.D., and Knier, B.L. 1988. Substituted o-iodoso- and o-iodoxybenzoic acids: synthesis and catalytic activity in the hydrolysis of active phosphorus esters and related systems. Journal of Organic Chemistry, 53: 3972–3978.

59) Gray, P.J. and Dawson, R. 1987. Kinetic constants for the inhibition of eel and rabbit brain acetylcholinesterase by some organophosphates and carbamates of military significance. Toxicol. Appl. Pharmacol., 91:140–144.

60) Maxwell, D. and Doctor, B.P. 1992. Enzymes as pretreatment drugs for organophosphate toxicity. In, Chemical Warfare Agents, ed. S. Somani. New York: Academic Press, pp. 195–207.

61) Koelle, G.B. 1963. Cholinesterases and anticholinesterase agents. Handbuch Der Experimentellen Pharmakologie XV, Berlin, Germany: Springer-Verlang.

62) Sidell, F.R. 1992. Clinical considerations in nerve agent intoxication. In, Chemical Warfare Agents, ed.S. Somani. New York: Academic Press, pp. 155–194.

63) Somani, S.M., Solana, R.P. and Dube, S.N. 1992. Toxicodynamics of nerve agents. In, Chemical Warfare Agents, ed. S. Somani. New York: Academic Press, Inc., pp. 67–123.

64) Marrs, T.C., Maynard, R.L. and Sidell, F.R. 1996. Organophosphorus nerve agents. In, Chemical Warfare Agents-Toxicology and Treatment. New York: Wiley and Sons, pp. 83–100.

65) Taylor, P. 2001. Anticholinesterase agents. In, Goodman and Gilman's The Pharmacological Basis of Therapeutics (10th Edn.), eds. J.G. Hardman, L.E. Limbird and A.G. Gilman. New York: McGraw- Hill, pp. 175–191.

66) Farahat, T.M., Abdelrasoul, G.M., Amr, M.M., Shebl, M.M., Fareahat, F.M. and Anger, W.K. 2004. Neurobehavioral effects among workers occupationally exposed to organophosphorous pesticides. Occup. Environ. Med., 60:279–286.

67) Roldan-Tapia, L., Parron, T. and Sanchez-Santed, F. 2005. Neuropsychological effects of long-term exposure to organophosphate pesticides. Neurotoxicol. Teratol., 27:259–266.

68) Rothlein, J., Rohlman, D., Lasarev, M., Phillips, J., Muniz, J. and McCauley, L. 2006. Organophosphate pesticide exposure and neurobehavioral performance in agricultural and non-agricultural Hispanic workers. Environ. Health Perspect., 114:691–696.

69) Rohlman, D.S., Lasarev, M., Anger, W.K., Scherer, J., Stupfel, J. and McCauley, L. 2006. Neurobehavioral performance of adult and adolescent agricultural workers. Neurotoxicology. 2006 Dec 1.

70) Eckerman, D.A., Gimenes, L.S., de Souza, R.C., Galvao, P.R., Sarcinelli, P.N. and Chrisman, J.R. 2007. Agerelated effects of pesticide exposure on neurobehavioral performance of adolescent farm workers in Brazil. Neurotoxicol. Teratol., 29:164–175.

71) Bazylewicz-Walczak, B., Majczakowa, W. and Szymczak, M. 1999. Behavioral effects of occupational exposure to organophosphate pesticides in female greenhouse planting workers. Neurotoxicology,\ 20:819–826.

72) Benschop, H.P., Trap, H., Spruit, H., van der Wiel, H., Langenberg, J.P. and DeJong, L.P.A. 1998. Low-level nose-only exposure to the nerve agent soman: atropinized guinea pigs. Toxicol. Appl. Pharmacol., 152:179–185.

73) Riddle, J.R., Brown, M., Smith, T.C., Ritchie, E., Brix, K. and Romano, J.A. 2003. Chemical warfare and the Gulf War: a review of the impact on Gulf War veterans' health. Mil. Med., 168:606–613.

74) Atchison, C.R., Sheridan, R.E., Duniho, S.M. and Shih, T.-M. 2004. Development of a guinea pig model for low-dose, long-term exposure to organophosphorus nerve agents. Toxicol. Mech. Methods, 14:183–194.

75) Klaasen, C.D. and Eaton, D.L. 1991. Principles of toxicology. In, Casarett and Doull's Toxicology: The Basic Science of Poisons (4th Edn.), eds. M. Amdur, J. Doull and C.D. Klaasen. New York: Pergamon Press, p. 12–49.

76) Hulet, S.W., McDonough, J.H. and Shih, T.-M. 2002. The dose–response effects of repeated subacute sarin exposure on guinea pigs Pharmacol. Biochem. Behav., 72:835–845.

77) Shin, L.M., Rauch, S.L., Pitman, R.K. 2006. Amygdala, medial prefrontal cortex, and hippocampal function in PTSD. Ann. N. Y. Acad. Sci., 1071:67–79.

78) Langston, J.L., Adkins, A.L., Moran, A.V., Rockwood, G.A. and Deford, M.S. 2005. Effects of sarin on the operant behavior of guinea pigs. Neurotoxicol. Teratol., 27:841–853.

79) Sipos, M.L., Smrcka, V.L., Zinkand, S.E., Kahler, D.W., Moran, A.V. and Atchison, C.R. 2001. Effects of subacute low dose exposure to sarin and soman on the acoustic startle response in guinea pigs. Soc. Neurosci. Abstr., 27: No. 959.17.

80) Thomson, S.A. et al. 2004. Low Level Chemical Warfare Agent Toxicology Research Program—FY03 Report and Analysis, Report number AFRL-HE-WP-TR-2004–0011.

81) Thomson, S.A. et al. 2005. Low Level Chemical Warfare Agent Toxicology Research Program—FY04 Report and Analysis, Report number AFRL-HE-WP-TR-2005–0054.

www.ingramcontent.com/pod-product-compliance
Lightning Source LLC
Chambersburg PA
CBHW080948170526
45158CB00008B/2411